研究生"十四五"规划精品系列教材

Quality Graduate Teaching Materials for the 14th Five-Year Plan of Xi'an Jiaotong University

International Journal Article Writing and Conference Presentation (Science and Engineering)

国际期刊论文写作与会议交流（理工类）

主　编　史文霞　陈　琦
副主编　郑四方
编　者　（姓氏笔画排列）
　　　　田荣昌　史文霞　李蓓岚　杨　蕾　陈　琦
　　　　邵　娟　郑四方　姜冬蕾　铁　瑛

西安交通大学出版社
XI'AN JIAOTONG UNIVERSITY PRESS

图书在版编目（CIP）数据

国际期刊论文写作与会议交流. 理工类 : 汉文、英文 / 史文霞，陈琦主编. -- 西安 : 西安交通大学出版社，2024.4
ISBN 978-7-5693-3734-1

Ⅰ.①国… Ⅱ.①史…②陈… Ⅲ.①理科（教育）-论文-写作-汉、英②工科（教育）-论文-写作-汉、英③理科（教育）-国际学术会议-学术交流-汉、英④工科（教育）-国际学术会议-学术交流-汉、英 Ⅳ.①G312②G321.5

中国国家版本馆CIP数据核字（2024）第082696号

国际期刊论文写作与会议交流（理工类）

主　　编	史文霞　陈　琦
副 主 编	郑四方
责任编辑	牛瑞鑫
数字编辑	宋庆庆
责任校对	李　蕊
装帧设计	伍　胜

出版发行	西安交通大学出版社 （西安市兴庆南路1号　邮政编码710048）
网　　址	http://www.xjtupress.com
电　　话	（029）82668357 82667874（市场营销中心） （029）82668315（总编办）
传　　真	（029）82668280
印　　刷	陕西博文印务有限责任公司
开　　本	787 mm×1092 mm　1/16　印张 16　字数 414千字
版次印次	2024年4月第1版　2024年4月第1次印刷
书　　号	ISBN 978-7-5693-3734-1
定　　价	58.00元

如发现印装质量问题，请与本社市场营销中心联系。
订购热线：（029）82665248　（029）82667874
投稿热线：（029）82665371

版权所有　侵权必究

前　言
Preface

　　英语已经成为国际科技领域的主流语言，国际学术期刊和国际会议通常采用英语作为工作语言，各国学者通过在国际学术期刊上发表英文学术论文实现相互交流、学习和推广科技成果的目的。国际期刊论文发表和国际会议展示是各国科研工作者的一项重要工作。研究生作为科技队伍的新生力量，应具备学术英语写作、发表与汇报展示的能力，这些能力是其科研素质的重要组成部分，也是高校培养国际型、创新型人才的重要衡量指标之一。近年来国内研究生学术写作和会议交流的教学活动取得了长足的发展，已形成了以体裁分析法为理论指导的主流教学模式，据此编写的教材也应运而生。然而，目前的写作教学与教材较多关注学科论文的共性，较少关注学科间的差异，教材中的例文选择更倾向于较易被理解的社会性或公共性话题，不能有效地突出不同学科论文的语篇特点。实际上，不同学科的论文存在较大差异，只强调共性而忽略特性无法有效对学生进行指导，甚至可能产生误导。目前，虽然已有一些针对特定学科领域的学术论文写作教材，但数量有限，仍无法满足市场需求。"国际期刊论文写作与会议交流"系列教材正是在这样的背景下编写而成的。本套教材包括理工、医学和社会科学三个分册，教材基于三大领域内的真实期刊论文和国际会议资料，充分展现了三大类学科论文的结构和语言特点，基于功能教学法、语类分析、跨文化交际、语域分析和"脚手架"等理论框架，对各领域学生有针对性地进行写作指导。在编写本套教材的过程中，我们有机融合了思想政治教育的元素。通过深入讨论学术论文的写作规范，培养学生严谨细致的科研态度，加强其对学术诚信的认识。此外，教材中选取了我国科研人员在国际顶尖期刊上发表的论文作为案例进行分析，这不仅有助于学生深入理解学术写作的常规，还能在潜移默化中增强学生的民族自豪感、国家自豪感和自信心。本套教材既可供高校教师、研究生（博士生、硕士生）、高年级本科生及相关科研人员学习使用，也可用于对国际学术会议参会人员的培训。

　　本册为理工分册，依据学术写作教学的经典理论——体裁教学理论，将解构、建模、再认、模仿共建和独立建构的教学思想融入教材编写，实现教学方法从以教师为主体向以教师为主导、学生为主体、产出为目的的转变，搭建"脚手架"，以用促学，学以致用。选用的主要素材为近年来发表在SSCI一区或二区期刊上的论文，涵盖能源动力、材料化学、电子信息、计算机、水资源工程、地理遥感等领域的研究，涉及医工结合、能源与环境工程结合、地理与环境工程结合等跨学科、跨领域研究。按照理工大类期刊论文的通用结构，即 IMRD（Introduction-Methods-Results-Discussion）安排教材内容，对论文结构及对应句法、时态、典型用语等语言特征进行分析并设置相应的写作练习。

本册教材共十个单元。第一单元和第二单元涉及学术论文写作的准备阶段，旨在让学生了解其专业领域内有影响力的期刊、文献的检索方法和引用方式、研究性论文的体裁特征等，培养学生的修辞意识。第三单元至第七单元主要介绍实证研究性论文主要组成部分的写作策略，包括引言、研究方法、研究结果、讨论、结论、标题、摘要等，侧重分析各部分的信息要素及典型语言特征。第八单元关注论文的写后修改。第九单元帮助学生了解期刊论文发表的流程和环节，涉及作者与期刊编辑及论文审稿人之间书信往来的措辞、语言的修改与润色等问题。第十单元的主题是国际会议交流，主要介绍口头汇报和海报展示等会议展示形式，从视觉呈现和口头呈现两个方面对学习者进行指导。

本教材的每个单元都设计了大量的实例和练习供教师和学生使用。在使用本教材时，建议教师可以进行下述操作。

1）在每个自然班内，可将学生按照专业接近原则建组，每组3至4人。

2）要求每位学生从自己专业的国际一流期刊中挑选3篇影响因子较高的论文作为个人的语料积累。论文必须是实证研究性论文，内容与自己的研究兴趣有关。

3）师生在课堂上共同分析教材例文的结构、信息要素和典型的语言特征。学生课后仔细研读和分析自己所选的3篇本专业期刊论文，确认论文的结构和信息要素是否和教材所述典型结构一致，如有差异，分析产生差异的原因，同时关注这些信息要素中的典型语言特征。

4）学生需比较自己选择的3篇论文在结构和语言方面的异同，加深对语言功能的理解。同一小组内，学生还需要对比不同期刊论文的研究范式和写作常规，分析不同期刊论文的共性和差异，以小组为单位在课堂上进行汇报和交流。

在本册教材编写的过程中，我们参阅并借鉴了大量国内外相关文献资料和同类教材，主要的参考教材和文献资料均已列在书后，在此向所有相关的作者表示深深的感谢！如需获取本书相关教学资料，请发送邮件至 speedtool@126.com。此外，我们还咨询了西安交通大学电子与信息学部李飞教授和西安交通大学电气学院贾少峰副教授，感谢两位专家帮助我们更加深入地了解工科领域的学术交流。西安交通大学出版社的编辑们也为本书的出版付出了大量的汗水和辛勤的劳动，在此一并向他们致以诚挚的谢意！本书虽经反复讨论和精心编写，但由于编者才疏学浅，加之时间仓促，书中不妥之处和谬误在所难免，恳请学界各位专家、学者及广大读者朋友提出宝贵的批评、修改意见和建议。

编　者

二〇二四年二月

目 录
Table of contents

Unit 1　Preparing for Your Writing　　/1

Unit 2　Understanding Journal Articles　　/19

Unit 3　Introducing Your Study　　/41

Unit 4　Describing Your Methods　　/77

Unit 5　Presenting Your Results　　/105

Unit 6　Discussing Your Study　　/137

Unit 7　Writing the Title and Abstract　　/171

Unit 8　Editing Your Paper　　/195

Unit 9　Submitting Your Paper　　/209

Unit 10　Presenting at Conferences　　/227

References　　/247

Sample Articles　　/248

Unit 1

Preparing for Your Writing

Learning objectives

In this unit, you will
- understand the purpose, audience, tone, and content of research writing;
- locate target journals for your reading and publishing;
- learn about tools for managing sources.

Self-evaluation

Recall an academic lecture you have listened to or a research article you have read recently, and discuss the following questions with your partners.
- When and where was the lecture offered? What was it about?
- Who was the speaker? Who attended the lecture?
- What was the goal of the lecture? How did the speaker achieve the goal?
- When and in which journal was the article published? What is it about?
- Why did you read the article? Did you obtain what you expected?
- What is the purpose of the article? How did the authors achieve the purpose?

Students who are trying to step into their disciplinary communities need to learn how to communicate appropriately with their peers. Human communication takes various forms and involves the use of various resources, such as verbal (spoken and written language), nonverbal (body language and instruments), visual (diagrams, images, and animation), and audio (sound and music). All these resources help realize the purposes of communication. To achieve effective communication, the speaker or writer needs to consider the rhetorical situation. The following questions may help you understand the rhetorical situation of a text.

- What is the topic?
- What is the purpose of the speech or writing?
- Who is the target audience?
- What tone is used in the speech or writing?

The meaning of a text may change as it may have been written for a specific audience, in a specific place, and during a specific time. These factors are considered the rhetorical situation of writing, a concept emphasizing that writing is a social activity, produced by people in particular situations for particular goals. Awareness of this concept can help writers and readers think through and determine why texts exist, what the writers aim to do through the texts, and how they may achieve their purposes in particular situations. Multiple factors are involved in rhetorical situations. In this unit, we will focus on the following important elements.

- **Audience:** the individual or group **whom** you intend to talk to.
- **Purpose:** the reason **why** you compose the text.
- **Context of use:** the place and situation **where** the text is used and interpreted.
- **Tone:** your **attitude** or **stance** conveyed through the text.

These elements are correlated with the content of your writing.

UNDERSTANDING RHETORICAL SITUATIONS

The table below contains details of a research project on rising sea levels. Imagine that you are writing for each of the following audiences: ① your supervisor/boss; ② scientists; ③ the general public; ④ politicians; ⑤ high school students. Discuss with your partner what detailed information would be interesting and relevant to a certain group of audience.

Categories of information on sea level rise	Interested audience
The dollar damage caused by sea level increases each year	
A literature review of previous research on rising sea levels	
Descriptions of calibration procedures for your instruments	
Some basic physics of how tides and currents work	
How much your project costs	
A log of all your measurements during the whole project	
A list of people who worked on the project	
Specifications of a new instrument to measure water conditions	
A new result showing a connection between sea level and coastal developments	
Procedures you used to avoid statistical biases in your data	
Your plans for further measurements	
Your recommendations for future research	

Rhetorical considerations in academic writing

Any piece of writing is produced to be read by some target audience, such as kids at 5 to 7 years, parents, college students, researchers in a specialized field, or common people. The writer always tries to achieve certain purposes through the text, for example, to express opinions or feelings, to provide information, to disseminate knowledge, to persuade, or to entertain. Where and how the text is used may also need to be considered as it will influence how the writer presents the text and in turn how the reader interprets it. Driven by the purpose and influenced by the topic, audience, and context of use, the writer may choose the corresponding content and appropriate tone to make the writing effective.

In universities, writing activities are performed as they are required for the completion of courses and degrees, for education purposes such as editing textbooks and class materials, and research purposes such as applying for research grants and publishing articles in scientific journals. Through various types of academic texts, authors are trying to achieve two major purposes: to inform and to persuade. A textbook, for example, is focused on informing readers of knowledge about a specific topic. A set of instructions or regulations in a lab is to inform the users of the steps to take or rules to follow when using the lab. A research article, however, informs the audience of a research project that has been completed and the findings obtained, as well as persuading the readers to believe that the work is important and valuable, and the findings are based on rigorous research procedures and thus reliable and valid.

Task 1.1 Read the texts below. Discuss with your partner the potential audience, purpose, context of use, and tone of the texts.

Text 1

In reflecting on and enjoying the scientific triumphs of the past millennium, we should also honour the origins of science and give thanks to the Greek scientists of the millennium before Christ. There is no earlier society in which one can identify science as distinct from technology, and where there was an objective attempt to explain the way the world works.

[Source: Wolpert, L. The well-spring. Nature 405, 887 (2000). https://doi.org/10.1038/35016166.]

Text 2

This year's Annual Meeting of the American Association of Physicists in Medicine will exploit a new online platform to enable a global audience of medical physicists to share ideas, interact with colleagues, and learn about innovative products. The virtual meeting will take place on 25th–29th July, providing five days of real-time scientific, professional and educational content, while an on-demand option is also available for registrants to catch up on anything they missed during those five days.

[Source: https://physicsworld.com/a/aapm-meeting-highlights-the-creativity-of-science-and-innovation/]

Text 3

Every flow has a characteristic (or typical) length associated with it. For example, for the flow of fluid within pipes, the pipe diameter is a characteristic length. Pipe flows include the flow of water in the pipes in our homes, the blood flow in our arteries and veins, and the airflow in our bronchial tree. They also involve pipe sizes that are not within our everyday experiences. Such examples include the flow of oil across Alaska through a four-foot diameter, 799-mile-long pipe, and at the other end of the size scale, the new area of interest involving flow in nano-scale pipes whose diameters are on the order of 10^{-8} nm.

[Source: Munson, B, Okiishi, T, et al. (2003) *Fundamentals of Fluid Mechanics*. John Wiley & Sons, Inc.]

Identifying the audience of research articles

Imagine you are going to give a presentation to an academic audience. Weeks before the big day, you spend time creating and rehearsing the presentation. You must make important, careful decisions about not only the content but also your delivery. Should you define or explain important terms, or will the potential audience already know them? Which part of your information would be most interesting to the audience? What questions do you anticipate that the audience may ask? Will you use technology to project figures and charts? Should you wear your suit and dress shirt? The answers to these questions will help you develop an appropriate relationship with your audience, making them more receptive to your message.

In this situation, the audience—the people who will watch and listen to the presentation—plays an important role. As you prepare the presentation, you need to profile the audience to anticipate their expectations and reactions. What you anticipate affects the information you choose to present and the way you will present it. Then, on the site of the presentation, you meet the audience in person and discover immediately how well your presentation addresses the audience.

This is also the case in writing. Although the audience—your readers—may not appear in person, they play an equally vital role. For research articles, there are multiple layers of audience.

- **Primary readers:** The primary readers should be those who read the paper after it is published in a journal. This group may include senior researchers who want to get updated about the cutting-edge development in the field and also graduate students who just start their academic careers and want to learn about their chosen fields and topics.

- **Secondary readers:** The secondary readers of your paper would be journal editors and reviewers. These people will evaluate your writing and decide whether the paper can be published.

- **Gatekeepers:** Your paper will also be read by your supervisor, the gatekeeper, who will look over your document before it is sent to other readers.

Each of these types of readers will look for different kinds of information. You may need to profile the readers in terms of demographics, education, prior knowledge, and expectations, based on which you can anticipate the needs of different readers when you draft your manuscript.

- **Demographics:** It is important to consider some basic information about your target audience, such as their age range, ethnicity, religious beliefs, or gender. Awareness of these factors helps you decide how you will adapt your writing to address the audience effectively.

- **Education:** Knowing the education level of the audience, you may alter your writing style, for example, being more formal or more casual.

- **Prior knowledge:** What the audience already knows about your topic will affect what you include in your writing. For example, you will judge whether or not to define terms and how detailed your explanation should be.

- **Expectations:** As mentioned above, readers will look for different kinds of information and have different expectations. You have to address the needs and expectations carefully to make your writing successful.

Task 1.2 In the scenario below, your paper will be read by various individuals. Discuss with your partner what each reader will look for and focus on while reading.

Scenario: You are preparing a review article and plan to seek feedback from your supervisor for revisions. Then, you will find an English teacher to polish the language. After that, you will submit it to a journal.

Selecting an appropriate tone in research articles

The tone of writing reveals the attitude and presupposition of the author. Just as speakers transmit emotions through voice, writers can transmit through writing a range of attitudes, from excited and humorous to somber and critical. These attitudes build a relationship between the audience, the author, and the text. To stimulate these connections, writers indicate their attitudes with useful devices, such as sentence structure, word choice, and punctuation. The writer's attitude must match the audience and purpose of writing appropriately. On the other hand, the tone of a text affects how the reader perceives the writer's intentions. Such perception, in turn, influences the reader's attitude toward the text and the writer.

In academic texts, the tone should reflect a writer's attitude, neutral and professional. Writers should inform and argue in an engaging but objective manner, and critique other people's works with courtesy and respect.

Writing with the right tone is critical, yet many students find it difficult to strike the right balance. It is easy to fall into the trap of writing in either a too formal or too casual tone. Both extremes can make your argument sound ill-researched and weaken the strength of your claim. To solve this problem, avoid writing in an overly formal tone which makes your paper sound as written by Shakespeare or the most esteemed professor. Worse than that, you may sound like you do not know what you are talking about. Meanwhile, make sure not to use colloquialisms—informal words or phrases—which make the writing sound less professional.

Exaggeration or hyperbole should be avoided. Remember, your reader is expecting proof of every statement you make. You may also be careful about making generalizations because those general statements may not be highly relevant and are hard to prove.

Task 1.3 Read the following paragraph and describe the writer's attitude toward wildlife conservation. Choose from the list of words provided those that you think suitable to characterize the writer's attitude.

A. Relaxed   B. Urgent   C. Bored   D. Impassioned   E. Funny   F. Well-informed

Many species of plants and animals are disappearing right before our eyes. If we do not act fast, it might be too late to save them. Human activities, including pollution, deforestation, hunting, and overpopulation, are devastating the natural environment. Without our help, many species will not survive long enough for our children to see them in the wild. Take the tiger, for example. Today, tigers occupy just 7 percent of their historical range, and many local populations are already extinct. Hunted for their beautiful pelt and other body parts, the tiger population has plummeted from 100,000 in 1920 to just a few thousand (Smith, 2013). Contact your local wildlife conservation society today to find out how you can stop this terrible destruction.

Task 1.4 Compare the two sentences in each pair and determine how the writer's tone varies between them.

Pair 1

S1: However, the researchers completely neglected the potential influence of the internal factors.

S2: However, the researchers did not address the potential influence of the internal factors.

Pair 2

S1: We wanted to see how effective biological filters are in reducing environmental impact.

S2: We sought to evaluate the efficacy of biological filters in reducing environmental impact.

Pair 3

S1: People with high levels of physical activity perform better on measures of attention than people with low levels.

S2: Many studies have found that people with high levels of physical activity perform better on measures of attention than people with low levels.

MANAGING YOUR SOURCES

 Search journals in your university e-library. Pick three journals in your field and try to find the information listed in the table below.

Information	Journal 1	Journal 2	Journal 3
Journal name			
SCI ranking			

(continued)

Information	Journal 1	Journal 2	Journal 3
Impact factors			
Journal scope			
Duration for processing a submitted paper			
Content requirements			
Acceptance rate			

Selecting target journals

Selecting the proper journal for your manuscript will increase the chance of getting published. The journal choice determines the size of the audience accessing and using your work and your professional reputation resulting from the publication. The right choice will help optimize the publication speed, accrue your professional reputation, and gain access to your desired audience.

A range of factors need to be considered when you choose where to publish. To start with, try to find peer-reviewed journals. The **peer-review** process helps establish the quality of your work and in turn, the quality of the journal is warranted. There is no easy way to assess the quality of a journal. Common measures of journal quality include **impact factor** (of one year or five years, showing the average citation counts of articles in a given journal within a period of time) and **journal ranking** (assessing the reputation and impact of a journal). Such information can be found either on the journal homepage or the website of a third party, such as *SCImago Journal Rank (SJR)*, a publicly available portal that includes the journals and country scientific indicators developed from the information contained in the Scopus® database.

It is of supreme importance that you find journals that publish articles on topics similar to yours. To know this, you can read the statements of the journal's aim and scope (available on the journal homepage) to see what topics the journal covers, or you can scan the titles of articles published in the journal in a recent couple of years to see whether your work would be of interest to the journal.

Sometimes, you may not know which journal to choose in face of so many journals in your field. Here is a tip. Look at the articles you have read about your topic, particularly those you plan to cite or have cited in your paper. Where are they published? The journals that appear more frequently in your list of references will be most likely to accept your work. In fact, by following back through the literature you should be able to develop a mind map of the journals in the field of your research. You can then check the websites or issues of these journals to identify those with scopes and aims most appropriate for your manuscript.

You may want to publish your paper quickly to ensure that similar works from other researchers would not be published before you. The average time needed by a journal to go through the initial checking, peer review procedure, and response to authors varies, but the average duration is around 12 weeks. If time to publication is critical to you, you should also check articles in the recent issues to find information about when a manuscript was received, revised, and accepted, based on which you will get an idea about the average time needed from submission to publication.

You may also want to know whether you need to pay for publishing your paper. Some journals charge fees, which could be of fixed amounts or based on the number of pages or whether there are colour illustrations. Many Journals offer to provide Open Access to your paper if you pay an upfront fee. Open access means your paper is accessible for free download without a subscription to the journal, which will make your work accessible to a wider range of readers. Check whether the journal of your choice offers this service or not if you want to pay for open access.

Task 1.5 Follow the steps to choose the right journals.

Step 1: Choose a potential list of journals in your field by considering the journal ranking, impact factor, and recommendation from your supervisor.

Step 2: Check the recent issues (maybe within the past two years) of the top journals on the list. Select the journals that publish articles on topics similar to yours.

Step 3: Choose the top three journals of your interest. Search the journal websites and complete the table below based on our discussion in this section.

Information	Journal 1	Journal 2	Journal 3
Journal name			
Impact factor / SCI ranking			
The time span for authors to get a response			
The average duration from submission to publication			
Open access (Y/N)			

Using academic research engines

When you start your research, one of the early steps is finding and reading the scientific literature related to your project. Your advisors are a great resource for recommendations about which past works, reported in journal papers, are critical for you to read. In addition to this, you'll need to be proactive and hunt for papers on your own. You may think about using regular search engines, but the results will not be satisfactory. You will get a wide mixture of websites, yet very few will be links to peer-reviewed scientific papers. To find reliable scientific sources, you should use an academic search engine.

There are many different academic search engines, some focused on a single field and others containing sources from multiple disciplines. There are a handful of free, publicly available academic search engines that can be accessed online; some of these are listed in Table 1.1. The majority is subscription-based. Universities and colleges often subscribe to academic search engines, so you can use them through your university library.

Table 1.1 Some free, publicly available academic search engines

Academic Search Engine	URL	Disciplines
Science Direct	http://www.sciencedirect.com/science/search	All
Web of Science	www.webofscienc.com	All
Pubmed	www.ncbi.nlm.nih.gov/pubmed	Life science
IEEE Xplore	Ieeexplore.ieee.org/Xplore/guesthome.jsp	Electronic and electrical engineering/Computer science
National Agriculture Library	Agricoda.nat.usda.gov	Agriculture
Education Resources Information Center	Eric.ed.gov	Education

When you begin your literature search, try several different **keywords**, both alone and in combination. As you view the results, you can narrow your focus and figure out which keywords best describe the intended topic. You can also do additional searches using the terms you learn from papers you have just found.

The results of academic search engines may come in the form of a title and abstract, together with the article information such as author names, journal title, volume, page numbers, year, etc., based on which you can find the full text of the paper in the publisher website. Some search engines, like PubMed, do provide links to free online versions of papers.

Task 1.6 Try to find the following article and explore the journal where it is published. Discuss with your partner the questions below.

Wan S, Jiang H, Guo Z, et al. Machine learning-assisted design of flow fields for redox flow batteries. *Energy Environ. Sci.*, 2022,15, 2874–2888.

1. How did you find the article? Did you use your university library? What academic search engine did you use?
2. Is the full text of this article available? How can you get the full text?
3. What have you learned about the journal? Would it be a potential target journal for your paper? Why or why not?

Searching Literature in Web of Science

The Web of Science is the world's most popular publisher-independent global citation database. As a comprehensive platform, it allows you to track ideas across disciplines and time from almost 1.9 billion cited references accross more than 171 million records. It includes SCI (Science Citation Index), SSCI (Social Science Citation Index) and A&HCI (Arts & Humanities Citation Index). Over 9,000 leading academic, corporate and government institutions and millions of researchers trust the Web of Science to produce high-quality research, gain insights and make more informed decisions that guide the future of their institution and research strategy.

Below are the key steps and some tips for using this platform.

Formulating your research question

Try to formulate your research question. Think critically about what you want to find out, for example: what is known about substance abuse in teenagers?

Deciding the searchable components

Now you need to decide the keywords used to find articles. You can use the Building Blocks method, in which you extract the core concepts from your question and then list as many synonyms as possible for each core concept. Take the research question above for example. The key concepts in the question are teenagers and substance abuse. The synonyms for these concepts could be adolescents, youth, young adults, high school students, children, young people, addicts, substance abuse, addiction, alcohol abuse, drug abuse, addictive behaviour, cocaine, etc.

Using Boolean operators to construct a search strategy

After identifying the core concepts in the question and determining synonyms for each concept, you should combine them logically to improve the relevance of your search results. Such operators as **AND** or **OR** can be used to construct search sets. OR will search for articles containing any of the terms you choose. Use OR to combine synonyms, alternative spellings or related items. AND will search for articles which contain all of the terms you have chosen. For example,

OR: Combining terms for the same concept e.g. teenagers OR adolescents OR youth
AND: Combining different concepts e.g. teenagers AND substance abuse

You can also use brackets to group concepts and force an order of operations. For example, you could input the query, (teenagers OR adolescents OR youth) AND (substance abuse OR addiction OR alcohol abuse). In addition, you may want to broaden your search to include plurals, grammatical variations and spelling variations, so you can use wildcards including asterisk (*), question mark(?) and dollar sign ($). The asterisk (*) represents any group of characters, including no character (e.g. s*food will find seafood and soyfood). The question mark (?) represents any single character (e.g. wom?n will find women and woman). The dollar sign ($) represents zero or one character (e.g. isch$emia will find ischaemia and ischemia). $ can be placed in the middle or at the end of the word. These wildcard tools can also be combined to get the broadest possible variation (e.g. organi?ation* will find organisation, organization, organisations, organizations, organizational, organizational). To search for an exact phrase, you enter it in quotes, e.g. "drug abuse".

It is noted that you usually have to carry out more than one search, as the initial search often generates either too few or too many results. You may also find that too many of the articles are not relevant to your topic. In such a case, you are suggested to return to your research question and make the necessary changes to your search.

Task 1.7 Explore how to use the Web of Science for a literature search. Do the following.

Step 1: Conceive a potential research question.
Step 2: Extract the core concepts of the question.
Step 3: Find the synonyms for the core concepts.
Step 4: Try to search with the keywords decided in the preceding step.
Step 5: Try advanced search by using wildcards.
Step 6: Compare the results obtained in Steps 4 and 5.

Using reference managing software

What are citation tools?
Citation tools help you store, organize, and share your research citations. They also automatically format your bibliographies and in-text citations into whatever style you need (APA, MLA, Chicago, and many more).

> **Which tool should I use?**
>
> The widely used citation managers include EndNote Basic, Mendeley, RefWorks, and Zotero. Talk with people in your department about the tools they use and why. You can always shift from one tool to another, as sources can easily be transferred between tools (Note that attachments do not always transfer.).

Task 1.8 Explore the citation tools available. Discuss with your partner which one you will use and why.

Task 1.9 Explore how to use EndNote (or another tool of your choice). Share your experience with your partner.

Unit task

Analyzing Potential Target Journals

To optimise the outcomes of publishing your manuscript, we recommend that you develop a publishing strategy. Part of the publication strategy is selecting your manuscript's target journal. Do the following to finish the task.

Step 1: Select three or four preferred journals in your field that you think would accept your manuscript. Scan the recent issues of the journal to see if there are articles on a topic similar to yours. If there are no such articles, consider choosing another journal.

Step 2: Find out information about the journals, such as impact factor, ranking, charge, open access, how fast an author will get a response, and how long it will take for a paper to be published after acceptance. Meanwhile, consult your professors or peer students about the journals' reputation. With all the information considered, decide whether the journals are good targets and which one you would give priority to.

Step 3: Use Web of Science or other databases to search articles published in the journals of your choice in the past 5 years. Find those relevant to your research.

Step 4: Choose a reference managing software, such as EndNote, to manage the articles you have found.

Unit 2

Understanding Journal Articles

Learning objectives

In this unit, you will
- gain a general understanding of different types of journal articles;
- explore the structural features of original research articles (RAs);
- get familiar with some common language features shared by original RAs.

Self-evaluation

Read Sample Article 1 (SA1) "Machchine Learning-Assisted Design of Flow Fields for Redox Flow Batteries" and answer the following questions.
- What information can I obtain from the abstract?
- What did the researchers explore?
- Which part should I read if I want to know why they conducted the study?
- Where can I find the details of their research procedures?
- Where can I find the announcement of their findings?
- Where do the authors relate their results to others' findings?
- Where may I find comments on the value or implications of the study?

Various types of articles are published in academic journals, each performing a unique function and all serving for knowledge dissemination, information exchange, and researcher conversation. Understanding these article types may help readers and writers better decide what to read and what to write, and importantly, where to publish. This is particularly important to novice researchers (e.g., graduate students) who are in urgent need of understanding the communicative conventions in their particular disciplinary communities.

This unit will help you explore the types of articles published in scientific journals. You will get to know what each type of article is and who would be the potential writers or target readers. Furthermore, you can think about what article types you will choose to read or write. Then, this unit will focus on a special type—articles that report original research, a major category of articles that researchers seek to read and publish. You will learn about the structural and linguistic features of such articles.

TYPES OF ARTICLES

Search online the official websites of four top scientific journals. Read through one of the recent issues of each journal to see what types of articles are published. Complete the table below with the information you have found.

Journal	Issue No.	Category / Article type
Science		
Nature		
PNAS *Proceedings of the National Academy of Science*		
J. Chem. Phys. *Journal of Chemical Physics*		

Task 2.1   All the four journals listed in the table below publish articles reporting original research, but they vary in naming/categorizing the articles and manuscript requirements. Search for relevant information on the journal websites and complete the table below.

Journal	Category	Description	Manuscript Requirements
Science			
Nature			
PNAS			
J. Chem. Phys.			

Task 2.2   Search the homepages of the four journals listed in the table of Task 2.1 and answer the following questions.

1. What types of articles does the journal publish?
2. How is each article type described or defined?

Task 2.3   Research article and review article are two types that you are likely to write. It would help if you do some work before putting pen to paper. Write down the possible work in the table below.

Article type	What you need do before writing
Review article	
Research article	

Unit 2   Understanding Journal Articles   21

STRUCTURAL FEATURES OF RAs

 The first six Sample Articles (SA1–SA6) are published in different journals. Read the articles and complete the table below.

Article	In what journal is it published?	What sections are included?	Which sections contain subsections?
SA1			
SA2			
SA3			
SA4			
SA5			
SA6			

Macrostructure of RAs

Original research articles (RAs) typically consist of the following components:
- a **title** that highlights the research;
- an **abstract** that provides a brief but comprehensive summary of the research;
- an **introduction** that sets the research in a context (or gives the "big picture") by providing a review of related past studies and developing research questions or hypotheses;
- description of **materials and methods**, which provides details of how the research was conducted, including the research design, study area, subjects, materials and tools, data, theories, algorithms, models, equipment, investigative procedures, data analysis methods, etc.;
- presentation of study **results**, including the summaries and interpretations of the measures obtained from investigative procedures;
- **discussion** over the outcomes of the measures as well as the overall study;
- **conclusions** that finalize the report.

Introduction (**I**), methods (**M**), results (**R**), and discussion (**D**) are considered four essential components of an RA's body. The contents of the components may be organized variably to form different sections which are titled differently, but these components represent the basic information to understand what the research is about, why the research was conducted, how the research was carried out, what results were obtained, how the reported work would influence others, and what needs to be done in the future. In brief, the four components form the conventional macrostructure of an RA, namely, the **IMRD** format. Many published papers adopt this structure but some may be still organized slightly differently. Fig. 2.1 shows the conventional macrostructures of RAs.

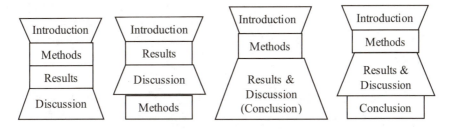

Figure 2.1   The conventional macrostructures of RAs

The four components are typically organized in the given sequence. However, in some journals or disciplines, the methods component is placed at the end of the article or attached as supplementary materials. Such placement is not left to the discretion of the writers but is specified in the guidelines for authors of the journals. There are, of course, variations of the structure. Sample Articles 1 and 5 (SA1 and SA5) have an integrated *Results and Discussion* section. This [**RD**] pattern is common and even prevails in certain fields.

Despite the wide existence of the combined [**RD**] section, for pedagogical reasons, we decide to analyze and teach the features of stand-alone **R** and **D** sections. It is important for novice writers to understand and distinguish the distinct functions and structural-linguistic features of the two components even if they ultimately write in the combined [**RD**] structure.

Conclusion (**C**) is traditionally the very last part of discussion in the conventional **IMRD** structure. In recent years, it is increasingly set apart and stands as an independent section to signal explicitly the completion of an article. This independence makes this component more salient, but its functions still overlap largely with those of discussion. For pedagogical simplicity, we do not analyze it as a separate section.

In brief, this textbook follows the conventional **IMRD** structure to analyze and teach how to write RAs. However, teachers and students need to be aware of the variations of article structure. For future research writers, a good strategy is to always check the specific requirements and the drafting trends in your target journal.

Task 2.4 Find three journals in your field. Scan the original research articles published recently to see how they are structured. Pick three RAs from each journal and complete the table below.

Journal	Article	Title/Topic	Section Heading
	1		
	2		
	3		
	1		
	2		
	3		
	1		
	2		
	3		

Task 2.5 Answer the following questions based on the information you have found in Task 2.4.

1. Are the RAs in the three journals following the IMRD structure? Are there any RAs in a varied structure, such as the IM[RD]? Are the RAs in the same journal following the same structure?

2. Is the *Methods* section always placed after the *Introduction* section? Do you know any journals or have you ever seen any articles where the *Methods* section is not following the *Introduction* section?

3. Is the study *conclusion* arranged as a separate section or the last part of the *Discussion section*?

4. Compare the section headings within the articles. Are they the same? Are there any differences between journals? Is it a good idea to replace the conventional headings with some new terms?

Meso-structure of RAs

Following disciplinary conventions, each RA section contains specific information performing expected functions. The details in each section need to be organized carefully so that the information flows in a logical, clear, and coherent manner.

The *Introduction* section is usually composed of paragraphs, whereas details in other sections are commonly organized into subsections and labelled with headings. The multiple layers of headings make the articles easy to scan. They may reflect the organization of contents within a section and the interconnection among contents across sections. Sample Article 5 (SA5) sets a good example.

2. Materials and Methods

2.1 Samples

2.2 Hot water decontamination and charge regeneration

2.3 Static electricity test

2.4. Waterproof test

2.5. Filterability test

3. Results and discussion

3.1. Decontamination of masks

3.2. Variation of electrostatic quantity

3.3. Waterproof property and microstructure analysis

3.4. Effects of decontamination on mask filterability

3.5. Effects of actual service processing

3.6. Practical application

In this paper, the details in the *Methods* section are organized into different subsections according to the important stages of experimental procedures, through which some results are obtained and reported in the corresponding subsections of the *Results & Discussion* section.

Task 2.6 Below are some (sub)section headings in Sample Articles 1 and 2 (SA1 and SA2). Underline the words which indicate actions or procedures performed in the research. Pay attention to the grammatical structures of the headings.

SA1	SA2

3. Results and discussion
    3.1 Library generation
    3.2 Multi-physics simulation
    3.3 Machine learning
    3.4 Screening and discovery
    3.5 Experimental validation
    3.6 Data exploration
	4. Materials and methods
	    4.1 Study area
	    4.2 Data
	    4.3 The processing steps of the GIS-Economic model
	        4.3.1. Identification of criteria
	        4.3.2. Determination of suitable locations for wind turbine installations
	        4.3.3. Estimation of the optimal purchasing price for electricity produced by wind turbines

Task 2.7 Read Sample Article 3 (SA3). Complete the following table with the subsection headings in the articles. Discuss with your partner how the contents in different sections are organized and how they are related.

Materials and Methods	
Results	
Discussion	

LANGUAGE FEATURES OF RAs

 Compare the two paragraphs below and discuss with your partner which paragraph is more academic and how they are different.

Paragraph 1

Using biosolids in agriculture can significantly improve crop growth and yield. When you add any organic matter to soil, you can improve the physical properties of the soil so that water can penetrate. In addition, the soil can become more porous and increase its bulk density, and the aggregates can become more stable. Furthermore, biosolids make the soil better able to absorb and store moisture.

Paragraph 2

The agricultural use of biosolids (with appropriate application rates) has been shown to produce significant improvements in crop growth and yield. The addition of any organic matter to soil improves the physical properties of the soil in terms of water penetration, porosity, bulk density and aggregate stability, and increases the soil's ability to absorb and store moisture.

Scholarly writings bear some characteristic lexico-grammatical features, such as nominalization, long and complex sentences, passive structures, and formal words. It is these features that make the text more academic. In this unit, we will look at some language features common to the entire article. Those specific of a particular section will be discussed in the subsequent units that elaborate on individual RA sections.

Nominalization

Science writers often need to deal with conceptual (abstract) ideas. They tend to isolate actions such as "investigating," "measuring," and "analyzing" as abstract conceptual units. This results in the use of nominal structures to represent abstract concepts entailing actions or changes. Such nominal structures center on the nouns derived from verbs or adjectives, which usually end with abstract noun suffixes such as *-tion*, *-sion*, *-ment*, *-ence*, and *-ness*. Below are some examples.

Original verb	Derived noun	Original adjective	Derived noun
determine	determination	productive	productivity
converge	convergence	durable	durability
measure	measurement	valid	validation
remove	removal	transparent	transparency
analyze	analysis	responsive	responsiveness

Reasons for the use of nominalization:

a. Nouns represent concepts, things, or people, often standing as technical or semi-technical terms, for example:

research participants   transparency of the film
soil contamination   accuracy in measurement
system performance   emergence of lithium batteries
signal transduction   detection of noises

b. The nominalized structures are used instead of clauses to make sentences more concise. This also avoids the mention of action-doers and helps create a more objective and "distanced" voice. For example:

Clausal structure	*If sound waves are compared with water waves, it can be found that they have something in common.*
Nominalization	*A comparison of sound waves with water waves shows that they have something in common.*
Action verbs	*In other words, by transforming verbs into their noun equivalents, writers can create complex sentences that are traditionally expected in academic writing. This process is called "nominalization," a term first coined by John Williams in his book Style: Toward Clarity and Grace.*
Nominalization	*First coined by John Williams in his book Style: Toward Clarity and Grace, "nominalization" is the term used to describe the transformation of a verb into a noun, thereby creating a complex sentence that satisfies the traditional expectations of academic writing.*

c. The nominalized structure serves as a summary of a previously mentioned idea. This helps condense the text and improve coherence. For example:

- Participants discussed the issues that had raised wide attention recently. The discussion lasted for several hours but no consensus was achieved.
- XXX was examined by … YYY was detected with … ZZZ was tested using … All these examinations were performed in triplicates.

d. Nominalization allows writers to demonstrate their powers of analysis by using descriptive words within the noun phrases. For example:

Importance	*the significant/remarkable/evident increase in …*
Speed	*the rapid/quick/gradual/slow increase in …*
Size	*the great/huge/small/minor/slight/marginal increase in …*
Trend	*the continuous/stable/sustained increase in …*

Reasons for avoiding nominalization:

Nominalized structures often represent abstract conceptual ideas that cause more mental burdens for readers. They also make the text sound less "active." Some experts suggest that abstract words should be avoided when concrete words can be used. In the following pairs of sentences, the ones using action verbs are more straightforward and easier to follow than those using nominalized structures.

Nominalization	*The expectation of the professor is for paper publication.*
Action verbs	*The professor expects you to publish.*
Nominalization	*Timing measurements were taken using a stopwatch.*
Action verbs	*Time was measured with a stopwatch.*
	A stopwatch was used to measure time.

In brief, science writing is known for its lexically condensed sentences choked with nominalized structures. There are good reasons for using such structures, but nominalization may not always be the best way to communicate. You need to make your decision by considering whether it will make your writing more effective.

Task 2.8 Rewrite the following sentences using nominalization to achieve conciseness and accuracy.

1. A recent report about road safety found people who drive too fast were the primary cause of accidents.
2. If the equipment is exposed for a long time, it will rapidly deteriorate.
3. Crops grow quickly because the soil in this area is very rich.
4. People may commit crimes because they are addicted to drugs.
5. The number of people in the world who eat GM foods every year has dramatically increased.

Task 2.9 The following sentences contain nominalization. Analyze them by following 3 steps.

Step 1: Underline the nominalized structures.

Step 2: Rewrite the sentence by replacing some nouns or noun phrases with verbs or adjectives.

Step 3: Compare the rewritten sentence with the original, decide which is better, and explain why.

1. The evaluation of this index was carried out by means of X.
2. The separation of the carbon of the nanotube's inner layer was demonstrated in two regions.
3. The evaluation of participants' performance was conducted by two researchers whose judgments were recorded on a computer.
4. Therefore, in this study, for the first time, OWA has been used to determine suitable locations for the construction of wind farms.
5. The observation of the extensive microstructural developments to the CeSiC coaxial nanotubes before and after ion irradiation were performed by TEM (Model 2100F, JEOL Ltd., Akishima, Japan) operating at 200 kv.
6. The preparation of polyamic acid (PAA) was executed by solution polycondensation between TPE-Q and BPDA in DMAc at room temperature.

Self-mention

In research writing, mentioning the researchers themselves or the reported study is realized by using the first-person pronouns (especially "*we*" and "*our*") or phrases such as "*this study,*" "*this work,*" "*the present study,*" "*this paper,*" etc.

Guidelines for the use of personal pronouns in research writing vary with discipline. A traditional assumption is that personal pronouns are sparingly used in hard sciences. Supposedly, they are avoided to maintain an objective tone, to keep focus on the materials or experimental actions rather than the researchers, and to keep a distance between the authors and the findings.

However, this assumption does not necessarily hold true. Scanning the six *Sample Articles*, you will find that the majority of the articles contain the subject pronoun "*we*" and the possessive pronoun "*our*," and most articles also use the phrases "*this study*" or "*the present study*" for self-mention. For instance, "*we*" appears over 20 times in SA1 and around 15 times in SA6; "*our*" appears nearly 20 times in SA6; "*this study*" appears 10-20 times in SA2, SA3, and SA4.

Undoubtedly, self-mention is inevitable when you write your paper. You can use either personal pronouns or impersonal phrases. That would be your own choice. What is important is to understand the proper context of using these words and phrases.

Task 2.10 **Find all the sentences containing "*we*" in Sample Articles 1 and 6 (SA1 and SA6). Write down the "*we* verb" phrases in the table below. Some examples are already given. Think about the following questions.**

1. Which section tends to have more sentences containing "*we*"? Why?
2. What is the verb tense in each case and why is the tense used? Can you identify some patterns in the use of tense in different sections?
3. Can you replace the verb with another one in each sentence?

Article	Section	we + verb
SA1	Introduction	we present we first develop
	Experimental	
	Results and discussion	
	Conclusions	
SA6	Introduction	
	Materials and methods	
	Results and discussion	
	Conclusions	

Task 2.11 Find all the sentences containing "*our*" in Sample Articles 1 and 6 (SA1 and SA6). Write down the "*our* noun" phrases in the table below. Some examples are already given. Discuss with your partner whether there are some relationships between the sections and the choice of the nouns.

Article	Section	our + noun
SA1	Introduction	
	Experimental	
	Results and discussion	our algorithm
	Conclusions	
SA6	Introduction	
	Materials and methods	
	Results and discussion	
	Conclusions	

Unit 2 Understanding Journal Articles   33

Task 2.12 Find the sentences containing "*this study*" in Sample Articles 2, 3, and 4 (SA2, SA3, and SA4). Read the sentences carefully to understand how the phrase is used. Think about whether it is good to change the phrase into "*we*" or "*our study*".

Long sentences

Scholarly writing is characterized by long sentences with complex structures, as authors need to deal with dense information and complicated ideas. Long sentences usually contain multiple clauses representing layers of information. The clauses can be connected with coordinating conjunctions (e.g., *and*, *but*, *so*, *for*, etc.) or subordinating conjunctions (e.g., *which*, *that*, *when*, *where*, *since*, *if*, etc.); they may also take various forms, diversifying the sentence structures.

One of the ways to vary sentence structure is to change a clause to a phrase, and vice versa. See the example below.

Clause	*As illustrated in Fig. 1, we first develop an end-to-end approach to designing flow fields, which encompasses library generation, multi-physics simulation, machine learning, and experimental validation, which will be presented in Sections 3.1–3.5 in order.* (SA1)
Phrase	*As illustrated in Fig. 1, an end-to-end approach to designing flow fields is first developed, encompassing library generation, multi-physics simulation, machine learning, and experimental validation, which will be presented in Sections 3.1–3.5 in order.*

The present participle "*encompassing*" is used to replace "*which encompasses*". By doing this, the subordinate clausal structure is removed and you do not need to worry about the verb tense. The sentence structure is simplified.

Task 2.13 The excerpt below is from Sample Article 1 (SA1). Discuss the following questions with your partner.

1. Which sentences are long?
2. How are the long sentences structured?
3. Would it be a good idea to break the long sentences into shorter ones?

> **1** Developing large-scale energy storage systems is an effective strategy to mitigate power fluctuations of electric grids with a high proportion of renewable energy sources (e.g., solar and wind).[1-3] **2** As one of the most promising energy storage technologies, the redox flow battery (RFB) is attracting increasing attention due to its decoupled energy and power, flexible scalability, fast response, and high safety.[4-6] **3** Among various RFB systems, the all-vanadium redox flow battery (VRFB) is the most widely studied and commercialized as it relies on vanadium ions as both positive and negative electroactive species, which significantly extends the cycle life of the battery by alleviating cross-contamination.[7-9] **4** However, the high capital cost poses a major barrier to the widespread application of the VRFB.[10-12] **5** An effective strategy for cost reduction is to develop VRFBs with higher power density.[13-15] **6** Over the past decades, tremendous progress has been made in increasing reaction sites and improving reaction kinetics of electrodes by surface modification, such as heteroatom doping, catalyst deposition, and surface etching, which are aimed at reducing the activation loss of the battery.[16-18] **7** Another important way to enhance the power density of VRFBs is to decrease the concentration loss by improving electrolyte distribution within the electrode, which can be realized by the design of flow fields.[19-21]

Task 2.14 Try to change the *which* clause in each sentence to a phrase. Compare the new sentence with the original and explain which one is better.

1. The under-rib convection can facilitate the penetration of electrolytes into the porous electrode, which improves the transport of electroactive species within the electrode and thus alleviates the concentration loss.

2. The Persian Gulf and the Gulf of Oman act as primary routes for access to open waters, which facilitate communication with other countries of the world.

3. Compared to the European and American rivers, Chinese rivers usually have a high SPM content and low water clarity, which negatively impact phytoplankton growth.

4. In addition, Chinese rivers are commonly characterized by turbidity, which adversely affects algal growth due to light limitation.

5. At the same time, radiology experts noticed that the manifestations of COVID-19 cases as seen through CT imaging had their own characteristics, which differed from those of the CT imaging manifestations of other viral pneumonias.

6. During the mask shortage, most people—including front-line medical staff—have been spontaneously reusing masks several times, which eases the tension between supply and demand to a certain extent.

7. Their use, however, makes it difficult to maintain generator operation at a reduced voltage, which causes subsequent disconnection of the source according to the Low Voltage Ride Through (LVRT) characteristic.

Check your understanding

Task 2.15 Choose two journal articles in your field. Scan the articles and complete the table below while considering the following questions.

1. How are the articles organized? Do they follow the IMRD structure?
2. Are the two articles organized the same way? Are the sections in the articles named similarly or differently?
3. Which sections are further divided into subsections?
4. How are the contents in different sections interrelated?

	Section headings	Subsection headings
Article 1		

	Section headings	Subsection headings
Article 2		

Task 2.16 Scan the two articles in Task 2.15. Do the following.

1. Find some sentences containing nominalized structures. Complete the table below.

Reasons for using nominalization	Examples
To represent concepts, things, or people, often standing as technical or semi-technical terms	1. 2.
To replace clauses to make sentences more concise	1. 2.

Unit 2　Understanding Journal Articles

Reasons for using nominalization	Examples
To summarize a previously mentioned idea	1. 2.

2. Search the cases of self-mention in the articles. Complete the table below.

	Self-mention	No. of occurrence	Examples
Article 1	we		
	our		
	this study		
Article 2	we		
	our		
	this study		

3. Select one paragraph with a moderate length from each article. Complete the table below.

Article 1	No. of sentences	
	Length of the longest sentence (words/sentence)	
	Length of the shortest sentence (words/sentence)	
	Average sentence length (words/sentence)	
Article 2	No. of sentences	
	Length of the longest sentence (words/sentence)	
	Length of the shortest sentence (words/sentence)	
	Average sentence length (words/sentence)	

4. Choose two super long sentences from the articles and try to break them into shorter sentences.

   Sentence 1:

   Sentence 2:

Reading Articles Related to Your Topic

A very important part of your study and research is to read about other people's works. Simply put, you need to read a lot of journal articles in your field, particularly those related to the topic(s) of your interest. Do the following to finish this task.

Step 1: Choosing articles

- Locate 10–20 published journal articles that are highly relevant to your research topic/interest.
- Pick 3 from them and download the full text from the university e-library.

Step 2: Reading for ideas

- Scan the section headings in an article to understand its overall structure.
- Locate and highlight important or useful ideas with highlighters (of different colors). Make notes where necessary.

Step 3: Reading for language

- Highlight terminologies / topic-specific expressions that you need to know. Compare them among the three articles.
- Highlight useful phrases and sentence patterns that you may use in your own writing.
- Share with your partner what you have learned from the articles.

Unit 3

Introducing Your Study

Learning objectives

In this unit, you will
- understand the general function and purpose of the *Introduction* section;
- learn about the common information elements in the *Introduction* section;
- develop the linguistic strategies for writing an effective *Introduction* section.

Self-evaluation

Read through the *Introduction* section of Sample Article 5 (SA5) "Can Mask Be Reused After Hot Water" and answer the following questions.
- Does this section provide the information you expect?
- What is the focus of each paragraph?
- Why is the topic worth researching?
- What motivated the researchers to conduct this study?
- What did the study aim to do?

The *Introduction* section, as the very first part of a paper, is the right place for readers to set the scene and make a good first impression of the work reported. It serves as a transition by moving readers from the world outside the paper to the world within. It should provide a brief overview of the research topic, give reasons for conducting the research, and inform readers of what to read in the following sections.

The *Introduction* section is of vital importance because it unfolds the paper for the readers and motivates them to continue reading. To some extent, a carefully crafted *Introduction* section acts as a springboard, establishing the direction for writing the entire paper. Readers will be more inclined to read the paper if the *Introduction* section is clear-cut, well-organized, and engaging. Journal editors, as the primary audience, will probably scan this section for evidence to answer the following questions.

- Does the work contribute something new?
- Is the contribution significant?
- Is it suitable for publication in the journal?

Writing the *Introduction* section can be slow and difficult. Some experienced authors recommend writing this section after finishing all other sections. This is because it may be difficult to figure out what exactly to put in this section until they have seen the big picture, felt confident about the chosen subject area, and backed up the arguments with appropriate references. Nevertheless, writing a good introduction section is of great importance. This unit will help you understand and grasp how to write an acceptable introduction section.

INFORMATION CONVENTION

Below is the *Introduction* section of Sample Article 4 (SA4). Read it and specify what each paragraph is about.

A Deep Learning System to Screen Novel Coronavirus Disease 2019 Pneumonia

Introduction

1 **1** At the end of 2019, the coronavirus disease 2019 (COVID-19) was reported [1-4]. **2** On 24 January 2020, Huang et al. [5] summarized the clinical characteristics of 41 patients with COVID-19, indicating that the common onset symptoms were fever, cough, myalgia, or fatigue. **3** All 41 of these patients had pneumonia, and their chest computed tomography (CT) examination showed abnormalities. Complications included acute respiratory distress syndrome, acute heart injury, and secondary infections. **4** Thirteen (31.7%) patients were admitted to the intensive care unit (ICU), and six (14.6%) died. Chan et al. [6] at the University of Hong Kong, China found evidence of human-to-human transmission of COVID-19 for the first time.

2 **5** The Diagnosis and Treatment Protocol for Novel Coronavirus Pneumonia (Trial Version 7) [7] recommended the etiological confirmation of patients with COVID-19 by means of two technologies: nucleic acid testing or specific antibody testing. The accessibility of nucleic acid testing has greatly improved in the past three months, but shortcomings still exist, such as high operating requirements, a time-consuming procedure, and a relatively low positive rate [8,9]. **6** A study from Wuhan showed that the positive rate of nucleic acid testing from oral swabs, anal swabs, and the blood of infected patients was 53.3%, 26.7%, and 40.0% respectively [10]. **7** Furthermore, antibody detection is not appropriate for early screening, since there is a window phase for antibodies testing. **8** Long et al. [11] reported that the median time of seroconversion for both immunoglobulin G (IgG) and IgM was about 13 d post onset.

Coronavirus 冠状病毒

onset 发病

Complications 并发症
respiratory distress syndrom 呼吸窘迫综合征

Etiological 病因学的

oral swabs 口腔拭子
anal swabs 直肠拭子

seroconversion 清转化
Immunoglobulin 免疫球蛋白血

3　**9** At the same time, radiology experts noticed that the manifestations of COVID-19 cases as seen through CT imaging had their own characteristics, which differed from those of the CT imaging manifestations of other viral pneumonias, such as influenza-A viral pneumonia (IAVP), as shown in Fig. 1. **10** Therefore, clinical doctors chose to replace nucleic acid testing with lung CT imaging as one of the early diagnostic criteria for this new type of pneumonia [12], with the aim of immediately curbing transmission.

radiology
放射学

4　**11** With the rapid development of computer technology, digital image processing technology has been widely applied in the medical field, including organ segmentation and image enhancement and repair, thereby providing support for subsequent medical diagnosis [13,14]. **12** Deep learning technologies, such as the convolutional neural network (CNN) with its strong ability of nonlinear modeling, have also been applied extensively in medical image processing [15–18]. Relevant studies have been conducted on the diagnosis of pulmonary nodules [19], the classification of benign and malignant tumors [20,21], and pulmonary tuberculosis analysis and disease prediction [22–24] worldwide.

segmentation
分割

CNN
卷积神经网络

5　**13** In this study, multiple CNN models were used to classify CT image datasets and calculate the infection probability of COVID-19. **14** These findings might greatly assist in the early screening of patients with COVID-19.

Paragraph	Focus of the paragraph
1	
2	
3	
4	
5	

Overall structure of the *Introduction* section

The functions listed in the table above correspond to some basic components of the *Introduction* section. Although what is covered in an introduction section may vary from one discipline to another and from one journal to another, some basic components are shared by most research papers. Fig.3.1 presents the possible information elements (IEs) that might be included in an introduction section. These elements can be classified into three information chunks: establishing the context, identifying the research gap(s), and filling the gap(s). Among these IEs, some are obligatory and others are optional.

The *Introduction* section usually starts from a general subject area and moves on to a particular research field. Therefore, the overall structure of the section can be thought of as a funnel shape—the top presenting the broadest view and the rest narrowing down to the specific research problem. Such a structure responds to two kinds of competition: competition for inviting readers to read the rest of the paper and competition for creating a research space.

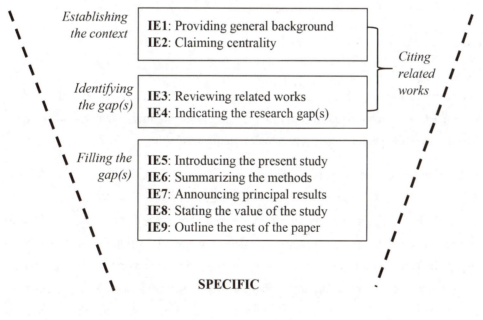

Figure 3.1 Information elements of the *Introduction* section

Not all the information elements may appear in a paper. The elements may not necessarily follow the given order. Some may appear in cycles. Some may be integrated and not clear-cut from each other.

Task 3.1 Read the *Introduction* section again and answer the questions below.

1. Does the structure of the *Introduction* section present a funnel shape? How does it move from general to specific?
2. Why does the author include references in Sentence 1?
3. Is it a good idea to omit Paragraph 1? And why?
4. Most research articles tend to begin by indicating that the chosen research field or topic is useful or significant. Please rewrite the first sentence into one addressing the importance of the research topic, the coronavirus disease 2019.
5. What is the research question this article tends to answer?
6. If you were invited to give the authors some revision suggestions, what other information would you expect to be covered in this section?

Establishing the context

At the beginning of a paper, it is common for authors to establish a context of the research to help readers understand how the research fits into a wider field of study. This information chunk contains two major elements.

IE1: Providing general background

In the *Introduction* section, authors mostly begin with broad statements that would generally be accepted as facts by the members of their target discourse community. In such statements, the present tense is often used because a function of this tense is expressing information perceived as always true. Sentences using the present perfect tense are also common, expressing what has been known or found over an extended period up to the present. When the statements contain more specific information, the sources of information will be indicated in the form of citations. Below are examples.

As an essential component of E-skin, pressure sensors mimic the human skin's perception of pressure. There are different types of sensory receptors with varied size, shape, number, and distribution to realize different characteristics of pressure perception in different areas of skin.

More than 80% of the world's energy production is driven by fossil fuels such as coal, oil, and natural gas [2,3]

IE2: Claiming centrality

Besides providing general information, the Introduction section tends to begin with an area, topic, or problem of wide interest to claim topic centrality, or to appeal to the discourse community. In this stage, authors may cite relevant works (if needed) to support their claims and connect the work to existing knowledge. Below are examples.

Electronic skins (E-skins) emulate the human sophisticated somatosensory system by transducing physiological signals into electrical signals and have great application potential in intelligence robots, biomimetic prosthetics, health monitoring, and various wearable devices.

Renewable energy is a crucial element of sustainable development [1].

To establish the context, you can undergo a process containing three steps: first, establish a "universe" for your readers; then, isolate one "galaxy" within this universe; and finally, lead your readers to one "star" in the galaxy (see Fig.3.2).

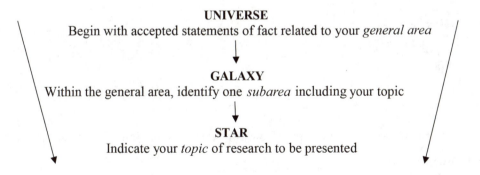

Figure 3.2   The logical order of establishing the context

Task 3.2 Below is part of the *Introduction* section of SA2. Read it and identify "the universe," "the galaxy," and "the star." Complete the table under the text. Meanwhile, underline the words or expressions indicating the importance of the research field or topic area.

> **The Site-Selection of Wind Energy Power Plant Using GIS-Malti-Criteria Evaluation from Economic Perspectives.**
>
> **1** Renewable energy is a crucial element of sustainable development [1]. **2** More than 80% of the world's energy production is driven by fossil fuels such as coal, oil, and natural gas [2,3]. **3** However, the main oil reserves on earth have been distributed unevenly, which can shortly lead to severe political and economic contentions [4]. **4** Increases in demand for energy and exhaustion of fossil fuels inevitably lead to higher prices. **5** As a result, most countries have developed policies related to renewable energies to reduce costs and become self-sufficient [5]. **6** Another incentive for using renewable energies to supply the energy demand is to be free of the environment's negative burden due to the consumption of fossil fuels [6].
>
> **7** The prospects of using renewable energies are considerably wide [7, 8]. **8** Some common renewable energy sources include solar, wind, geothermal, biomass, tidal power, hydrogen fuel, etc. **9** Wind energy is one of the fastest-growing renewable energy sources worldwide [9]. **10** Site selection is the first and most important step to establishing a wind power plant.

Universe	
Galaxy	
Star	

Task 3.3 Read the following part of the *Introductions* section of SA1 and answer the following questions.

> **Machine Learning-Assisted Design of Flow Fields for Redox Flow Batteries**
>
> **1** Developing large-scale energy storage systems is an effective strategy to mitigate power fluctuations of electric grids (电网) with a high proportion of renewable energy sources (e.g.,

solar and wind).[1-3] **2** As one of the most promising energy storage technologies, the redox flow battery (RFB) (氧化还原液流电池) is attracting increasing attention due to its decoupled(解耦的) energy and power, flexible scalability(可扩展性), fast response, and high safety.[4-6] **3** Among various RFB systems, the all-vanadium redox flow battery (VRFB) (全钒氧化还原液流电池) is the most widely studied and commercialized as it relies on vanadium ions as both positive and negative electroactive species(电活性物质), which significantly extends the cycle life of the battery by alleviating cross-contamination.[7-9] **4** However, the high capital cost poses a major barrier to the widespread application of the VRFB.[10-12] **5** An effective strategy for cost reduction is to develop VRFBs with higher power density.[13-15] **6** Over the past decades, tremendous progress has been made in increasing reaction sites and improving reaction kinetics of electrodes by surface modification, such as heteroatom doping(杂原子掺杂), catalyst deposition, and surface etching, which are aimed at reducing the activation loss of the battery.[16-18] **7** Another important way to enhance the power density of VRFBs is to decrease the concentration loss by improving electrolyte(电解质) distribution within the electrode, which can be realized by the design of flow fields.[19-21]

Questions:

1. Does the first sentence contain the keywords which could be found in the title?
2. Do the initial sentences introduce a general problem of wide interest? What is it?
3. How do the authors emphasize the necessity of addressing the general problem?
4. How do the authors highlight the importance of the research by using centrality claims?
5. Do the authors use references (e.g. citations) to support their centrality claims?
6. How does the *Introduction* section move from a general issue to a more specific topic?

Identifying the research gap(s)

The second information chunk is identifying the research gap(s), which consists of two elements: reviewing related works and pointing out the research gap(s). The latter is based on the former.

IE3: Reviewing related works

Reviewing previous works related to the topic is vitally important since it determines the objective(s), question(s) and methods of the research project at hand. When writing a literature review, you need to concisely identify the key works influencing or laying the groundwork for the research, trying not to miss important ones while not including those only loosely related or irrelevant.

There is a tendency among journal editors and reviewers to check the list of references first. The editors will do so to select peer reviewers. If an editor knows some particular reviewers who are experts in the field of your topic but whose classical works are not included in your list of references, the editor may not be convinced that you have done a good job in reviewing the literature. By the same token, a reviewer invited to read your paper as an important expert regarding the topic may not be happy if he or she finds "the paper does not cite me".

A good literature review should be organized logically (see Fig. 3.3). It is possible to arrange your citations in order from those most distantly related to your study to those most closely related. In addition, you may arrange the citations in chronological or thematic order. In a chronological review, the authors discuss previous studies in the order of the time when they were published, highlighting the changes over time concerning the specific issue. In a thematic review, the authors group and discuss previous studies in terms of the themes or topics they cover. The citations within each group can then be ordered chronologically or from the general to the specific.

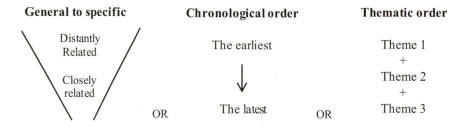

Figure 3.3 The logical order of reviewing related studies

IE4: Indicating the research gap(s)

After you have presented a contextual setting and discussed the related works of other researchers, you are supposed to identify the gap(s) to be filled by your work. The gap here means an unresearched or under-researched important research problem or area by other authors. By identifying the gap(s), you can not only present the reasons or motivations for your study but also make a clear and cogent argument that your work is important and possesses value. Usually, the statement of a research gap is brief, containing only one or two sentences.

To indicate a research gap, you can employ the following strategies.

Strategy 1: Stating something missed or limited in previous studies. Typical lexical signals include: however, nevertheless, yet, unfortunately, but; no/ little/ few/none previous research, fail, overlook, limited, etc.

Strategy 2: Stating something unclear in previous studies. This strategy points out an unanswered question or unsolved problem in prior studies, suggesting that additional research needs to be done to clarify the question or problem. It can be realized directly by *wh*-questions or indirectly by a statement about difficulty, disadvantage or problem.

Strategy 3: Stating something controversial in previous studies. This strategy points out conflicting findings or viewpoints regarding a specific topic in the current literature. These disagreements may be theoretical or methodological conflicts.

Strategy 4: Continuing a research trend or tradition. This strategy is employed when writers acknowledge that there have been related past studies in a specific domain, and use a relatively direct and overt statement to highlight the need to follow a recent trend or tradition, for instance, "Earlier studies seemed to suggest X. To verify this finding, more work is needed."

Strategy 5: Presenting positive justification. This strategy directly argues for the necessity of the present research.

It should be noted that IE3 and IE4 do not necessarily follow a linear sequence in which previous studies are thoroughly reviewed and then research gaps are identified. They are often presented in an alternating pattern in which they occur in pairs. No matter what pattern is used, research gaps should always be based on the literature review, narrowing the scope of the paper to the topic investigated in the present research.

Task 3.4 Read the following excerpts from the *Introduction* sections of research articles, and identify how the citations in each excerpt are organized. Determine whether they follow a chronological order, a thematical order, a general-to-specific order, or a combination of several orders.

Text A

A Framework for Suspect Face Retrieval Using Linguistic Descriptions

1 In recent years, many techniques for sketch-to-photo retrieval have been proposed. **2** Tang and Wang (2004) used eigenfaces to retrieve the photo from the sketches. **3** Liu, Tang, Jin, Lu, and Ma (2005) proposed a method to utilize the pseudo-sketch synthesis by using a locally linear embedding method with nonlinear discriminate analysis for sketch recognition. **4** Yuen and Man (2007) proposed a two-phase face retrieval system. **5** To match the suspect's mug-shot image with the sketch, they have utilized local and global feature measurement. **6** Recently, Wang and Tang (2009) used a multiscale Markov Random Fields (MRF) model to synthesize sketches automatically.

Text B

The Measurement of Mobility

1 The economic literature which discusses mobility and makes some attempt at measurement broadly falls into two categories. **2** In the first, elementary statistical techniques and indices such as the rank correlation coefficient are used to evaluate the changes in relative positions[6, 8, 11, 14, 19, 22]. **3** In the second category, measures of mobility are a by-product of simple stochastic specifications of changes over time [1, 9, 10].

Text C

On the Time Consistency of Optimal Policy in a Monetary Economy

1 The time-consistency issue is by no means a new one in economics. Strotz[25] appears to be the first one to have raised it in relation to an individual consumer. **2** More recently, however, Kydland and Prescott[15] have discovered a family of models exhibiting time inconsistency where the source of the problem lies in the technology and in the assumption that people hold rational expectations. **3** Although they briefly touch upon a monetary economy, the central results of their remarkable paper are given in a context where money plays no central role.

4 In the monetary literature, Auernheimer[2] appears to be the first one to have noticed that time inconsistency could arise if the government attempts to maximize the revenue from money creation.

Text D

Machine Learning-Assisted Design of Flow Fields for Redox Flow Batteries

1 Historically, the earlier VRFBs have been constructed using the flow-through configuration, where electrolytes enter a long channel packed with a thick porous electrode.[22] **2** This configuration, however, leads to an uneven distribution of electroactive species and high ohmic resistance, especially when scaling up the battery stack.[23] **3** The application of flow fields in the VRFB (the all-vanadium redox flow battery) has interested researchers since 2012. **4** Experimentally, Aaron et al.[24] reported a VRFB with a "zero-gap" serpentine flow field. **5** Their battery yielded a peak power density of 557 MW cm^2, much higher than the previous ones with the flow-through configuration. **6** Subsequent studies from different groups also demonstrated the superior performance of the VRFB with serpentine flow fields to that without flow fields.[25-27]

7 Generally, an ideal flow field for RFBs should maximize the uniformity of the reactant distribution over the electrode with a relatively low-pressure drop.[30-32] **8** To pursue this goal, some studies have been conducted to develop novel flow fields,[33-36] in which the spatial arrangement of flow channels is re-designed by combining expertise and intuition.

...

9 Machine learning-assisted screening, on the other hand, provides a promising avenue. **10** With the exponential growth of computing resources and improvements in simulation tools and machine learning techniques, this powerful approach has led to the discovery of novel materials and molecules across diverse applications[45-49] and has promoted advances in fluid mechanics.[69-72]

...

Task 3.5 The following sentences are from the *Introduction* section of a paper about food habits of undergraduate students. They are in scrambled order. Put them in the logical order by numbering them from 1 to 7.

_____ A. Young and Storvick (1970) surveyed the food habits of 595 college freshmen in Oregon and found that the men generally had better diets than the women.

_____ B. Litman et al. (1975) reported that green and yellow vegetables and liver (all nutritionally desirable foods) were not liked by teenagers in Minnesota public schools. They also found that teachers have almost no influence on their students' food habits.

_____ C. Studies of the food habits of young school children have shown that the diets of grade school children are often deficient in ascorbic acid, calcium and iron (Lantz et al., 1958; Patterson, 1966)

_____ D. A review of the literature indicates that food habit studies have been conducted with students from a variety of different age groups.

_____ E. Young (1965) examined the nutrition habits of a group of young school children and found that their mothers lacked information about the importance of milk and foods rich in ascorbic acid.

_____ F. Studies done with adolescent children report similar findings (Ohlson and Hart, 1970; van de Mark and Underwood, 1972).

_____ G. A number of studies have been conducted using both male and female college students as subjects.

Task 3.6 Read the following gap statements abridged from the *Introduction* sections of sample articles, and underline the signal words or expressions indicating a gap. Discuss with your partners what strategies are used to indicate it.

1. However, the existing national standards and local standards for masks mainly focus on the performance of disposable masks, and there are no specific requirements or instructions on mask performance for multiple uses, including the duration of a single use, disinfection methods, and the number of times a mask can safely be reused. How can masks be efficiently sanitized for multiple uses during the COVID-19 pandemic? Are masks damaged during decontamination treatment? How many times can masks be decontaminated by appropriate methods? In fact, there is no scientific theory or experimental data support to answer these questions.

2. However, previous estimations mostly used seasonally in situ data collected in European and American rivers (Dai et al., 2012b; Ludwig et al., 1996). Without long-term field data, previous studies only reported the spatial variations and/or short-term changes in OC transport in Asian rivers (Huang et al., 2017; Ran et al., 2013; Wang et al., 2012). Moreover, considerable attention was given to coastal rivers, but information is limited for

inland rivers (Huang et al., 2012; Ni et al., 2008). Therefore, the long-term OC transport features across different Asian rivers remain unclear. To better understand the global carbon cycle, there is an urgent need to determine the spatiotemporal changes of OC transport in different Asian rivers.

3. However, besides accounting for the limitations posed by national laws and regulations, selecting and assessing sites for renewable energy farms must also be considered from various technical, economic, social, and environmental perspectives.

4. Few studies were conducted to assess the PN emission characteristics of the different hybrid vehicles, and the main concern of the existing literature is mainly focused on the PN emission factor, while instantaneous PN emissions are seldom analyzed.

5. But their application to CSAs is problematic because of a large number of nonequivalent atomic configurations and, hence, atomic jump frequencies.

6. However, the objective functions used in the previous studies only consider maximizing the reaction rate but ignore minimizing the pressure drop, hence not reflecting the real problem behind the design of flow fields for RFBs. Unfortunately, even though successful in mechanical applications,[43] multi-objective topology optimization remains challenging for solving complex systems such as RFBs where multiple physicochemical mechanisms are coupled. In addition, the density approach used in the previous studies suffers from numerical instability (mesh dependency and local minima in particular[44]), which may result in large computational errors. More robust and effective methods to design flow fields for RFBs are required.

7. So far, very little attention has been paid to the effects of deposits on spray characteristics, combustion, and emission under different fuel temperatures and injection strategies. In addition, the interaction between the spray is essential to the GDI engine [34], and it is affected by the deposits, which is not fully characterized so far.

8. Recently, the effect of injection pressure on impinging ignition was studied by Du et al. [16]. They found that the ignition delay is longer with higher injection pressures when the fuel injection mass is fixed. However, in the spray-wall combustion experiments carried out by Shi et al. [17,36], it was found that in a low-temperature environment, the ignition delay gradually decreases with the increase of the injection pressure while the fuel injection mass is also fixed, but the combustion effect gradually worsens, and the misfire phenomenon

occurs at high injection pressures. Thus it can be seen that the injection pressure has a great influence on the spray-wall combustion, but the various laws are not clearly concluded and explained yet, the contradictions even appear in different pieces of literature, especially at low temperatures. In addition, a higher injection pressure can give better fuel atomization, but it also increases the chance of misfire under low-temperature wall impinging conditions. The explanation for the phenomena that higher injection pressure may cause poorer combustion or even misfire is very important to the diesel engine cold start, but no systematic numerical analysis has been done.

9. Although numerous researchers have analyzed the influences of DP mixtures on the emission and combustion performances of CI engines, there is a lack of research on the emission performances and combustion characteristics of a CI engine fueled with DBE/n-pentanol/diesel blends, particularly those that focus on the particle size distribution (PSD) differences at various EGR rates in a CI engine.

10. While much progress has been made, diffuse optical imaging methods continue to suffer from low spatial resolution due to the loss of high-frequency information as light propagates through scattering media [10].

Task 3.7 Read the following excerpts from the *Introduction* sections of SA1, and answer the questions below.

Machine Learning-Assisted Design of Flow Fields for Redox Flow Batteries

1 Machine learning-assisted screening, on the other hand, provides a promising avenue. **2** With the exponential growth of computing resources and improvements in simulation tools and machine learning techniques, this powerful approach has led to the discovery of novel materials and molecules across diverse applications[45–49] and has promoted advances in fluid mechanics.[69–72] **3** Compared with traditional high-throughput screening, it benefits from acceleration in data collection and is thus particularly cost-effective for solving complex problems. **4** However, using machine learning-assisted screening to design flow fields faces two challenges. **5** The first is how to generate a custom library containing numerous flow field designs. **6** Mathematically, the problem of generating flow fields can be formulated as the search for a Hamiltonian path in an undirected graph with a rectangular grid of cells, which has been proven to be NP-complete.[50] Similar problems have been investigated in the research area of labyrinth and maze generation.[51–53] **7** Due to inherent flaws and high computational cost, algorithms reported in previous studies are difficult to traverse the solution set; ergodicity is important for screening-based design approaches. **8** Therefore, a new algorithm that ensures the diversity of the generation result is needed. **9** The second challenge is how to screen promising candidates from the search library. **10** Though powerful and robust, machine learning-assisted screening remains restricted in terms of the inverse design with the goal of optimizing properties or performance objectives.[54] **11** To overcome this difficulty, some studies have combined machine learning with optimization algorithms (e.g., genetic algorithm) to tackle material design problems related to polymers,[55] battery electrodes,[56] and optical glass.[57] **12** Such methods, however, are infeasible when the solution set of the problem to be addressed is not continuous.

Questions:

1. How do the authors begin reviewing the previous studies on machine learning-assisted screening?
2. In which sentences do the authors point out the gap existing in the previous studies? What word signals the ending of the literature review and the beginning of a gap statement? What other words or expressions could also indicate this shift?
3. What strategies are used to point out the gaps?
4. Do the authors employ a citation while indicating a gap? Why is it used?
5. Why do the authors start sentence 11 with an infinitive phrase?
6. Do you think this paragraph is easy to understand? Why do you think so?
7. Which pattern is used to arrange the citations and the gap indication, sequential or alternating pattern?

Filling the gap

Reviewing related works contributes to the identification of a research gap that requires further work to fill. Naturally, the next step is to tell the readers what was done in this study as an effort to fill the gap. Several information elements can be included in this chunk, among which some are obligatory while others are optional. In this part, it is common for the authors to switch from the impersonal to the personal tone—by using the first person pronoun *we* or *our*—to make it clear what they intended to do or actually did in the present study.

IE5: Introducing the present study

This information element is essential and obligatory. You can introduce your study by stating the research aim explicitly or describing what is to be done or accomplished. Below is one example.

In this paper, we present the results of an extended atomistic study of vacancy diffusion in randomly disordered Ni-Fe alloys, which is one of the simplest face-centered cubic (fcc) solid solution model systems.

IE6: Summarizing the methods

Another highly frequent information element is to summarize the research methods. By doing so, you can give readers a sense of how you reached your conclusion. This also helps transition from the *Introduction* section to the next section which describes the research activities in detail. To make a summative description, you can only highlight the central or fundamental approaches or the most salient details. Below is an example.

To assess the instantaneous PN characteristics of the hybrid vehicles, four China-6 light-duty passenger hybrid electric vehicles with different technical parameters were tested under the real-world driving conditions following China-6 RDE regulation [6].

Apart from the above two major elements, some other possible elements may be included.

IE7: Announcing the principal results

IE7 helps readers preview the important findings of the study, which may encourage readers to continue reading through the paper. Note that not all the papers choose to disclose research results in this initial section. Different disciplines or journals may have their own preferences. Below is an example.

It was found that under a low frequency of around 1.3 Hz, an instantaneous peak power density of 10.98 W/m^3 and a transferred charge density of 0.65 mC/m^3 were achieved.

IE8: Stating the value of the study

IE8 tells readers how the findings in the study practically or theoretically contribute to the pre-existing body of research on the topic. Similar to IE7, this element is also a strategy for promotion. However, if you opt for a value claim, you need to be cautious or modest, not sounding overconfident or boastful. Below are examples.

The results are of great significance for understanding the global carbon cycle, balancing the carbon budget in inland waters, and managing river water quality.

One of the main contributions of this paper is our analysis of the SegNet decoding technique and the widely used Fully Convolutional Network (FCN) [2].

IE9: Outlining the rest of the paper

IE9 identifies the structure and content of the remaining parts of the paper, giving readers a sense of how information will be sequenced and presented as they read further. This element is more likely to be included in long papers with complex structures. It may not be necessary in short articles with fixed structures. Below is an example.

The remainder of the paper is organized as follows. In Section 2 we review related recent literature. We describe the SegNet architecture and its analysis in Section 3. In Section 4 we evaluate the performance of SegNet on outdoor and indoor scene datasets. This is followed by a general discussion regarding our approach with pointers to future work in Section 5. We conclude in Section 6.

These optional information elements appear in a small portion of articles. You may feel confused about which should be contained in the *Introduction* section. One useful guideline is to consider which information elements contribute to building the reader's understanding of and interest in the study. Another suggestion is to follow standard practice in your field.

Task 3.8 Read the following sentences carefully. Determine whether each of them is an explicit statement of the research purpose (P) or a description of what is to be done or accomplished (D). Underline the subject and the predicate verb of each sentence, and summarize the possible verb tense usage for introducing the present study.

1. Here, we present a framework for the machine learning assisted design of flow fields for RFBs.

2. The main objective of the present study was to provide a price estimation approach for wind energy-generated electricity using Geographic Information System-Multiplecriteria Evaluation (GIS-MCE) and economic models.

3. Hence, the goal of this study was to systematically investigate the influences of DBE, npentanol, and EGR on the performances, and PM and regular gas emissions of a CI engine.

4. In this study, multiple CNN models were used to classify CT image datasets and calculate the infection probability of COVID-19.

5. In this article, we summarize our experimental results and evaluations on three kinds of typical masks (disposable medical masks, surgical masks, KN95-grade masks) treated by the so-called "hot water decontamination + charge regeneration" approach.

6. The present study aims for a comprehensive study of the effects of the injector deposits on the characteristics of the spray, combustion, and emissions under different fuel temperatures, injection pressure, and injection strategies.

7. The purpose of this paper is to optimize the synergistic effect of EGR and hydrogen addition, reaching the ultimate goal which can not only reduce gaseous emissions.

8. Based on station-based monitoring and watershed data during 2004–2018, this study clarified the spatial patterns, temporal variations, and driving forces of OC transport in 41 rivers across China, the largest Asian country.

Task 3.9 Read the excerpt from the *Introduction* section of SA1 and fill in the table below the text.

> **Machine learning-Assisted Design of Flow Fields for Redox Flow Batteries**
>
> **1** Here, we present a framework for the machine learning-assisted design of flow fields for RFBs. **2** As illustrated in Fig. 1, we first develop an end-to-end approach to designing flow fields, encompassing library generation, multi-physics simulation, machine learning, and experimental validation, which will be presented in Sections 3.1–3.5 in order. **3** After exploring 11564 flow fields, the data-driven search process identifies eight promising candidates. **4** Three of these are fabricated and tested, and experimental results show that the battery with the newly designed flow fields yields exhibits about a 22% increase in limiting current density and up to 11% improvement in energy efficiency compared to the conventional serpentine flow field. **5** In Section 3.6, a statistical analysis of geometric properties shows that the promising candidates have the saved channel length of 1490+/–100 and the torque integral of 20.1+/–1.8, revealing the design rules of flow fields on a quantitative level for the first time.

Information element	Sentence No.	Tense
Introducing the present study		
Summarizing the methods		
Announcing principal results		

Task 3.10 Below are the last paragraphs of the *Introduction* sections of SA10 and SA11. Compare them with that in Task 3.9 and discuss with a partner how they are similar to or different from each other in terms of information elements and tenses used.

> **On the Existence and Origin of Sluggish Diffusion in Chemically Disordered Concentrated Alloys**
>
> **1** In this paper, we present results of an extended atomistic study of vacancy diffusion in randomly disordered Ni-Fe alloys, which is one of the simplest face-centered cubic (fcc) solid solution model systems. **2** This simplicity enables a clear identification of the underlying mechanisms of sluggish diffusion with little ambiguity. **3** By applying classical molecular dynamics (MD) on unprecedented μs timescales, combined with on-the-fly and

conventional kMC, we demonstrate that vacancy diffusion in concentrated alloys can be lower than diffusion in either of its pure components. **4** Moreover, we have identified the basic chemically dependent mechanisms that give rise to sluggish diffusion.

Blockchain Meets IoT: An Architecture for Scalable Acess Management in IoT

1 The present study aims for a comprehensive study on the effects of the injector deposits on the characteristics of the spray, combustion, and emissions under different fuel temperatures, injection pressure, and injection strategies. **2** The flow rate of the fouled injector and clean injector were firstly compared. **3** The Scanning Electron Microscopy (SEM) images of deposits were obtained to analyze the physical characteristics of deposits. **4** A high-speed camera and a long-working-distance microscope were used to study the spray characteristics of the fouled and a clean injector. **5** Then the combustion of the fouled injector was analyzed under different injection strategies in an optical GDI engine. **6** The diffusion flames in the combustion images were extracted and analyzed to assess the in-cylinder soot. **7** Meanwhile, a DMS500 was used to obtain the exhaust particle concentration; also, the DMS500 was used to verify the effectiveness of characterizing soot concentration using diffuse flame luminance.

Task 3.11 Analyze the sentences excerpted from the *Introduction* sections of research articles. Each sentence contains mutiple information elements. Discuss with your partner what information elements are present in each sentence. Additionally, underline the words or expressions that help you make your decisions.

1. This paper describes some demographic factors that might be important for a better understanding of rural-to-urban migration in developing countries.

2. The aim of this investigation was to study groundwater conditions in order to aid in evaluating the general hydrologic situation in the area under study.

3. We present a super-resolution localization imaging method that is an extension of earlier localization concepts [10], [16] and is based on spatiotemporal data.

4. The effect of soil temperature on Verticillium wilt disease in peppers is discussed in this paper, as is varietal susceptibility of the pepper host, both of which may be of importance in explaining the irregular occurrence of the disease in California.

5. To assess the instantaneous PN characteristics of the hybrid vehicles, four China-6 light-duty passenger hybrid electric vehicles with different technical parameters were tested under the real-world driving conditions following China-6 RDE regulation [6].

6. Therefore, in order to better understand the spray-wall combustion process under low ambient temperature conditions and explain the phenomena such as misfire at high injection pressure conditions, the wall-impinging flame characteristics at varied injection pressures with a large aperture nozzle (diameter 0.32 mm) were studied by using the numerical method.

Task 3.12 All the sentences below state research value. Underline the modal auxiliaries in the sentences and put them in the order from the most certain to the most tentative. Write down the corresponding sentence numbers in the blanks.

1. These findings might greatly assist in the early screening of patients with COVID-19.
2. This study would be a useful first step in the development of biodegradation techniques.
3. This categorization may contribute to understanding the landscape architecture design process and its communication and teaching.
4. This new method should simplify the calculating procedure.
5. With this method, the students will develop communicative skills effectively and express themselves creatively.
6. Such an approach can enhance the efficiency of the turbine and generator speed control and increase the speed of the primary frequency control.

Sentence number: _____ _____ _____ _____ _____ _____
Scale of tentativeness: Certain ⟶ Tentative

Task 3.13 Here are the sentences abridged from the *Introduction* section of a paper on using microcomputers in teaching. Work with a partner to put them back in the logical order by numbering them from 1 to 7.

_____ A. Many schools have purchased and are purchasing microcomputers for infusion into their directed learning programs.

_____ B. Although much work has been done to date, more studies need to be conducted to ascertain the effects of microcomputer-assisted instruction in teaching various subjects in a variety of learning situations.

_____ C. During the past 40 years, the U.S. has experienced the integration of the computer into society.

_____ D. Most individuals seem to agree that the microcomputer will continue to hold an important role in education (Gubser, 1980; Hinton,1981; Foster,1982).

_____ E. This topic is identified as being of importance to teachers in providing them with the necessary background to teach lessons in farm records.

_____ F. The purpose of this study was to ascertain the effect of using microcomputer-assisted instruction as compared to a lecture-discussion technique in teaching principles and methods of cost recovery and investment credit on agricultural assets to graduate students in agricultural education.

_____ G. Schmidt (1982) identified three types of micro-computer use in classrooms: the object of a course, a support tool, and a means of providing instruction.

Task 3.14   The paragraph below is an outline of a paper, whose sentence structures lack variety. Discuss with your partner what changes you would make to improve sentence variety.

1 The rest of the paper is organized as follows. **2** In Section 2, a literature review of ORR mechanisms and dispatching rules for assembly job shop scheduling will be given. **3** In Section 3, a description of the simulation settings will be given, such as the shop floor configuration, the assembly job information and performance measures of the job shop scheduling problem. **4** In Section 4, the ORR mechanisms and dispatching rules used in the job shop scheduling problem will be described. **5** In Section 5, the design of experiments will be given. **6** In Section 6, analysis of experiment results will be given. **7** In Section 7, conclusions and future research directions will be given.

LANGUAGE CONVENTION

Read the *Introduction* of SA4 again and answer the questions below.

1. Do all the citations follow the same format? How many different types of citations could you find?

2. What tenses are used in the sentences? Does any sentence use the tense that is out of your expectation? What is the possible reason for the authorial choice?

3. Do you think the *Introduction* section is easy to understand? How do the authors make it readable?

4. Suppose you were to cite the last two sentences in your paper, how do you paraphrase them?

Writing a definition

When writing about a topic, you need to clarify a term or concept which is unfamiliar to your readers or whose meaning is ambiguous or controversial. For this reason, it is common to write a definition in the *Introduction* section.

There are mainly three types of definitions, short definition, sentence definition and extended definition. A short definition or "gloss" gives information about a term in a word or phrase often placed within either parentheses, between dashes, or between commas. Sometimes, definitions are signalled by words or phrases such as *i.e., known as*, *defined as*, and *also called*.

A sentence definition is brief and somewhat similar to a dictionary definition. It mainly consists of three parts including the term to be defined, a word showing the category that the defined term belongs to, and some distinguishing characteristics of the defined term. The following example showcases the structure of a sentence definition.

Term	Be	Category word	Distinguishing characteristic
A solar cell	is	a device	that converts the energy of sunlight into electric energy.

An extended definition, longer than a short formal sentence definition, contains multiple sentences and provides a fuller explanation of the term. It usually begins with a general one-sentence definition, followed by additional details such as types, history, examples, components, or applications.

Task 3.15 Read the following definitions and underline the words or expressions signaling a definition.

1. PLA is a polymer obtained from corn and produced by the polymerization of lactide.

2. The uncertainty associated with the energy obtained from other types of non-utility generators, i.e., thermal and hydro, is relatively small compared to that associated with wind.

3. Based on the analysis of the whole process of VR (a scientific method and technology created by human beings to understand, stimulate, and better adapt and use the nature), this paper presents different categories of VR problems.

4. DOC refers to carbon bound in various organic humic substances, carbohydrates, proteins, viruses, etc. (Evans et al., 2002; Hope et al., 1994; Schulze, 2005).

5. Pfam, known as a comprehensive database of protein families, contains 7937 families in the current release.

6. Generally, a promising explanation for the improved properties of CSA is the so-called "sluggish diffusion".

Task 3.16 Read the following extended definitions of the term *thermometer*. Discuss with a partner how one sentence definition is extended into one paragraph.

1. A thermometer is an instrument which measures temperature. It consists of two important elements: a temperature sensor in which some physical change occurs with temperature, and some means of converting this physical change into a numerical value.

4. A thermometer is an instrument which measures temperature. It is widely used in industry to control and regulate processes, in the study of weather, in medicine, and in scientific research. For instance, a doctor uses a thermometer to measure a patient's body temperature; aircrafts use thermometers to determine if atmospheric icing conditions exist along their flight path.

3. A thermometer is an instrument which measures temperature. It is widely used in industry to control and regulate processes, in the study of weather, in medicine, and in scientific research. For instance, a doctor uses a thermometer to measure a patient's body temperature; aircrafts use thermometers to determine if atmospheric icing conditions exist along their flight path.

4. A thermometer is an instrument which measures temperature. In 1593, Galileo invented the first thermometer. In 1612, Santorio became the first inventor to put a numerical scale on his thermometer, but the instrument was not very accurate. In 1714, the first modern thermometer, the mercury thermometer was invented by Daniel Fahrenheit.

Task 3.17 Write an extended definition of an important term or concept in your research field.

The term or concept defined	
The first sentence	
Additional details	

Citing related works

Citing other published works is a prominent feature of research articles. It plays a critical role in constructing disciplinary knowledge. By doing so, the authors tell readers what has been done or known about the topic under investigation and how the present work fits into the entire system of research regarding the topic.

Different disciplines and journals have their preferred citation styles. You have to read the author guidelines offered by your target journal to see how you should format your citations and references. The in-text citation largely takes three formats. One is author names plus publication year; the second is a serial number (often as a superscript or within square brackets) indicating the location of the cited work in the list of references at the end of the paper; the third is author names plus a serial number.

According to the information focus salient in a citation, the following three types of citations are available, which perform different functions and bear different linguistic features.

Information-prominent citation

This type of citation focuses on the cited idea or information, whose specific source information is attached to the sentence in either of the in-text citation formats. The simple present tense is generally used because the cited idea or information is presented as facts.

Among them, the temperature, wetlands, and croplands are usually reported to increase riverine OC (Correll et al., 2001; Freeman et al., 2002; Tian et al., 2013).

In this context, the Manning classical theory of diffusion [20–22] and kinetic Monte Carlo (kMC) methods [23–26] are commonly used for predicting diffusion properties of concentrated solid solutions.

Author-prominent citation

This type of citation highlights the cited author whose name stands as a sentence component, followed by the publication year in parentheses or the serial number as a superscript or within square brackets. The simple past tense is conventionally used for this type of citation to emphasize a particular study.

Park et al. [7] carried out the experiments on the impinged spray by using a high-speed camera and phase Doppler particle analyzer system and found that the impingement led to batter atomization compared with the free spray.

Yaji et al.[41] reported a two-dimensional (2-D) topology optimization method to design a flow field with the maximized generation ratio of vanadium ions.

For seven U.S. coastal rivers, Tian et al. (2013) reported that a one-degree centigrade increase in air temperature would raise the riverine DOC concentration by 0.476 mg/L.

Another study by Noorollahi et al. [38] employed a combination of GIS and AHP to evaluate wind energy potential in Markazi Province, Iran.

> **Weak author-prominent citation**
>
> In this type of citation, the cited author's names do not appear in the sentence, but such nouns as *several studies, previous studies or some researchers* are used as a general reference to more than one author, which means several studies are used as the source. This type of citation emphasizes the cited idea or finding and thus tends to use the present perfect tense. Below are some examples.
>
> *To pursue this goal, some studies have been conducted to develop novel flow fields,*[33–36] *in which the spatial arrangement of flow channels is re-designed by combining expertise and intuition.*
>
> *Many studies have found that, compared to the conventional ICE vehicles, relatively high particle emissions are emitted from hybrid electric vehicles [5,40,41]*
>
> It is noted that even though certain tenses are typically favored for each type of citation, nowadays there is a tendency for more scientists to use the simple present tense in citing others' works.

Task 3.18 The sentences below are excerpted from the *Introduction* sections of SAs. Fill in each blank with the proper verb tense and explain the possible reason for your choice.

1. Song et al. [31] _____ (find) that deposits at different locations _____ (have) different effects on injection behaviors.

2. Without long-term field data, previous studies only _____ (report) the spatial variations and/or short-term changes in OC transport in Asian rivers (Huang et al., 2017; Ran et al., 2013; Wang et al., 2012).

3. Recently, the effect of injection pressure on impinging ignition _____ (study) by Du et al. [16]. They _____ (find) that the ignition delay _____ (be) longer with higher injection pressures when the fuel injection mass _____ (fix). However, in the spray-wall combustion experiments carried out by Shi et al. [17,36], it _____ (find) that in a low-temperature environment, the ignition delay gradually _____ (decrease) with the increase in the injection pressure while the fuel injection mass _____ (also fix), but the combustion effect gradually _____ (worsen), and the misfire phenomenon _____ (occur) at high injection pressures.

4. Yaji et al.[41] _____ (report) a two-dimensional (2-D) topology optimization method to design a flow field with the maximized generation ratio of vanadium ions. They then _____ (extend) the method to a three-dimensional (3-D) level.[42] Their simulation results _____ (suggest) that the VRFB with the optimized flow field _____ (deliver) better performance than those with conventional flow fields.

Task 3.19 Read the following excerpts from SA8 and SA9. Discuss with your partner the similarities and differences between them in citation type and verb tense.

SA8

A Numerical Investigation of Injection Pressure Effects on Wall-Impinging Ignition at Low-Temperatures for Heavy-Duty Diesel Engine

Many studies have investigated the effects of different factors on air–fuel mixing [7–12], ignition and combustion characteristics [13–17] under the impinging spray conditions. Park et al. [7] carried out the experiments on the impinged spray by using a high-speed camera and phase Doppler particle analyzer system and found that the impingement led to batter atomization compared with the free spray. They concluded that the radial penetration of impinging spray increased with the injection pressure and the injector-wall distances, but decreased with the ambient pressure. Zhang et al. [8] investigated the air–fuel mixture process after impingement with different injection pressure coupled with the nozzles of different diameters at an ambient temperature of 797 K (Tamb = 797 K) and found that the small diameter nozzle and high injection pressure are helpful to vaporization and air–fuel mixture at ultra-high injection pressures.

SA9

Characteristics of Instantaneous Particle Number (PN) Emissions from Hybrid Electric Vehicles under the Real-Word Driving Conditions

Many studies have found the significant gap between laboratory tests and real driving in both fuel consumption and pollutant emission [43]. Duarte et al. [44] found that for the conventional vehicles, the RDE fuel consumption was 23.9% and 16.3% higher than certification values of NEDC and WLTC respectively. For hybrid diesel vehicles, on-road CO_2 value is 52%–178% higher than certification [45]. Both the EU and China have taken the RDE test into vehicle certification [43].

Information flow

To establish a smooth flow of ideas in your writing, you can connect sentences by using an old-to-new information pattern. That is to say, you can start a sentence with the information mentioned previously and give new information in the later part of the sentence. Such an old-to-new pattern helps connect sentences while extending the text. There are several ways to indicate old information.

1. Using word repetition or derivation

*Approximately three years ago, an apparently new and unexplained disorder called acquired immune deficiency syndrome (**AIDS**) was recognized. Characteristically, **AIDS** is associated with a progressive depletion of T cells.*

*In order to assign data examples to a set of predefined categories, we need to classify data first. One prerequisite for any **data classification** is to have labelled examples.*

2. Using synonyms

*Human behavior has a large influence on the global ecology. Many of the environmental challenges facing us today are a direct result of **human actions**.*

3. Using pronouns or demonstrative pronouns

*Ice forms when water is cooled to 0°C and continues to lose heat. Generally, **this** happens when the air temperature falls below 0°C.*

*Water is one of the most intriguing substances on earth. **It** has the interesting property that its freezing point is within the range of the earth's surface temperature variation for significant parts of the year.*

4. Using "This/these + summary noun"

*It is also a huge mistake to not validate the user's identity. To alleviate **this problem**, we might provide security training to developers.*

International Journal Article Writing and Conference Presentation (Science and Engineering)

Task 3.20 The following text is not well written in terms of information flow. Please improve it by following two strategies: using old-to-new information pattern; moving from general information to more specific information.

1 Pleuropneumonia (APP) can present as a dramatic clinical disease or as a chronic, production limiting disease in pig herds. **2** A sudden increase in the number of sick and coughing pigs and a sharp rise in mortalities among grower/finisher pigs may herald an outbreak of APP in a herd. **3** On the other hand, signs may be limited to a drop in growth rate and an increase in grade two pleurisy lesions in slaughter pigs. **4** The disease surfaced in the Australian pig population during the first half of the 1980s and ten years later was regarded as one of the most costly and devastating diseases affecting the Australian pig industry.

Characteristic expressions

In the *Introduction* section, some characteristic expressions are used to perform the following function:
- Claiming centrality;
- Reviewing related works;
- Identifying the gap(s);
- Introducing the present study.

Scan the QR code for a list of the expressions. Try to get familiar with them and pick some to use in your future writing.

Task 3.21 Complete the following sentences by translating the Chinese in parentheses into English.

1. _____ （大规模储能系统开发的既往研究主要集中于）an effective strategy to mitigate power fluctuations of electric grids with a high proportion of renewable energy sources.

2. Unfortunately, _____ （即便在机械应用中取得了成功），

multi-objective topology optimization _____ (仍然具有挑战性) for solving complex systems.

3. The redox flow battery (RFB) _____ (一直是深入研究的主题) in the last decades.

4. _____ (已有研究未考虑到) how to screen promising candidates from the search library.

5. _____ (本研究的目的是开发) an end-to-end approach to the design of flow fields by combining machine learning and experimental methods.

6. _____ (为了解决这一难题), some studies have combined machine learning with optimization algorithms.

Check your understanding

Task 3.22 Choose two from the sample articles (SA1–SA11) and read the *Introduction* sections. Finish the tasks below.

1. Analyze the IEs included.
2. Locate the IEs not presented in detail and try to explain the reason.
3. Analyze the tenses and voices used.
4. Highlight characteristic expressions.

Task 3.23 Select two research papers in your field. Analyze the *Introduction* sections and answer the following questions by filling in the table below.

1. Which information elements are present or absent?
2. What tense(s) is/are used in each element?
3. Which information elements contain citations?
4. What types of citations are used?
5. Is the first pronoun "I" or "we" used to refer to the authors?

Information element	Present or not	Tense	Citation type	"I" or "We"
Providing general background				
Claiming centrality				
Reviewing related works				
Pointing out the research gap				
Introducing the present study				
Summarizing the methods				
Announcing principal results				
Stating the value of the study				
Outlining the rest of the article				

Unit task

Writing Your *Introduction* Section

Up to now, you have already learned how to search for studies related to your own research interest, how to manage retrieved literature, and how to write the *Introduction* section of a research article. It is time to set about drafting your own *Introduction* to your selected research topic. Do the following to finish this task.

Step 1: Choose 10–20 studies as references for your research proposal. List the references in the format required by your target journal.

Step 2: Compose an **OUTLINE** for the *Introduction* section. Indicate clearly the following.
- A general area related to your topic + known facts (*sources if necessary*)
- A subarea (*if there is one*) + known facts (*sources if necessary*)
- A specific area or your topic + previous studies (*references*)
- Centrality of the area / subarea / topic (*where appropriate*)
- Research gap(s)
- Research aim(s) / question(s) / overview
- A summary of research methods

Step 3: Draft the *Introduction* section. It should be about 500 words long, containing essential information elements and citations.

Unit 4

Describing Your Methods

Learning objectives

In this unit, you will
- understand the general function and purposes of the *Methods* section;
- learn about the common information elements in the *Methods* section;
- develop the linguistic strategies for writing an effective *Methods* section.

Self-evaluation

Read the *Methods* section of Sample Article 5 (SA5) "Can Masks Be Reused After Hot Water Decontamination During the COVID-19 Pandemic" and answer the following questions.
- How is this section named?
- Does this section provide the information you expect?
- How do the authors organize this section?
- Do the subheadings match with those in the *Results* section?
- What is the major verb tense used in this section? Why is it used?
- What is the major voice used in this section? Why is it used?

The *Methods* section (together with the *Results* section) provides the essential information of the reported study and is thus the core of writing the paper. This section describes in detail how the research was carried out. Its main purposes are to establish credibility for the reported work and to allow other researchers to repeat the procedures and obtain similar results. For both purposes, necessary and sufficient information about the work is needed.

The *Methods* section, compared with other sections, is often considered the easiest to write. The authors just need to describe the work they have done, so they are clear about what to write in this section. During writing, they can refer to other articles reporting similar works. What's more, the language in this section is rather formulaic. For these reasons, some authors write this section before other sections.

The *Methods* section usually follows the *Introduction* section and precedes the *Results* section, but in some journals, it is placed at the end of the article in smaller forms or even attached as supplementary information. The placement of this section is specified in each journal's Guidelines for Authors. However, this does not mean that this section is not important in paper writing. Instead, it is crucial to write a good *Methods* section since it not only undergirds the study but functions as a decisive factor for the "fate" of a manuscript. Journal editors often make critical judgments based on their reading of this section. This unit will help you understand and grasp how to write an acceptable *Methods* section in terms of structure and language.

INFORMATION CONVENTION

 Below is the *Methods* section of Sample Article 4 (SA4). Read it and specify what each paragraph is about.

A Deep Learning System to Screen Novel Coronavirus Disease 2019 Pneumonia

Materials and Methods

2.1. Study dataset

1 A total of 618 transverse-section CT samples were collected in this study, including 219 from 110 patients (mean age 50 years; 63 (57.3%) male patients) with COVID-19 from the First Affiliated Hospital, College of Medicine, Zhejiang University; Wenzhou Central Hospital; and the First People's Hospital of Wenling from 19 January to 14 February 2020. **2** All three hospitals are designated COVID-19 hospitals in Zhejiang Province. **3** Every COVID-19 patient was confirmed with real-time reverse transcription-polymerase chain reaction (RT-PCR) testing from sputum or nasopharyngeal swab, and cases with no image manifestations on the chest CT images were excluded. **4** In addition, there was a gap of at least two days between CT datasets scanned from the same patient in order to ensure the diversity of samples. **5** The remaining 399 CT samples were collected from the First Affiliated Hospital, College of Medicine, Zhejiang University as the controlled experiment group. Of these, 224 CT samples were from 224 patients (mean age 61 years; 156 (69.6%) male patients) with IAVP including H1N1, H3N2, H5N1, H7N9, and so forth; 175 CT samples (mean age 39 years; 97 (55.4%) male patients) were from healthy people. **6** The diagnosis of IAVP was proved by the RT-PCR detection of viral RNA from sputum or nasopharyngeal swab. **7** There were 198 (90.4%) COVID-19 and 196 (87.5%) IAVP cases in early or progressive stages; the remaining 9.6% and 12.5% cases, respectively, were in the severe stage; no significant differences in stages between two diseases. **8** IAVP CT samples were used because it was critical to distinguish IAVP from patients with suspected COVID-19 currently in China.

transverse 横向的

RT-PCR 逆转录聚合酶链反应
sputum 痰
nasopharyngeal swab 鼻咽拭子

IAVP 甲型流感病毒性肺炎

RNA 核糖核酸

2   **9** The ethics committee of the First Affiliated Hospital, College of Medicine, Zhejiang University approved this study, and the research was performed in accordance with relevant guidelines and regulations. **10** All participants and/or their legal guardians signed an informed consent form before the study took place.

3   **11** A total of 528 CT samples (85.4%) were used for training and validation sets, including 189 samples from patients with CCOVID-19 194 samples from patients with IAVP, and 145 samples from healthy people. **12** The remaining 90 CT sets (14.6%) were used as the test set, including 30 COVID-19 cases, 30 IAVP cases, and 30 healthy cases. **13** Furthermore, the test cases of the CT set were selected from people who had not been included in the training stage.

2.2. Process

4   **14** Fig. 2 shows the whole process of COVID-19 diagnostic report generation in this study. **15** First, the CT images were preprocessed to extract the effective pulmonary regions. **16** Second, a three-dimensional (3D) CNN segmentation model was used to "segment" multiple candidate image cubes. **17** The center image together with its two neighbors of each cube was collected for further steps. **18** Third, an image classification model was used to categorize all the image patches into three types: COVID-19, IAVP, and irrelevant to infection (ITI). **19** Image patches from the same cube "voted" for the type and confidence score of this candidate as a whole. **20** Finally, the overall analysis report for one CT sample was calculated using the Noisy-OR Bayesian function [25].

pulmonary 肺的

Paragraph	Focus of the paragraph
1	
2	
3	
4	

Overall structure of the *Methods* section

The *Methods* section has different headings such as *Methods*, *Materials* and *Methods*, *Experimental*, *Method Description and Validation*, *Simulation*, *Design*, *Model*, and so on. Although the name varies with disciplines, research fields and journal conventions, this section mainly describes how the study was carried out, what materials were adopted to accomplish the study, and why certain methodological choices or decisions were made. Therefore, some basic information elements are shared by most scientific research papers. These elements could be categorized into three information chunks: contextualizing study methods, describing study subjects and/or materials, and describing study procedures. Fig. 4.1 presents the common information elements (IEs) of the *Methods* section.

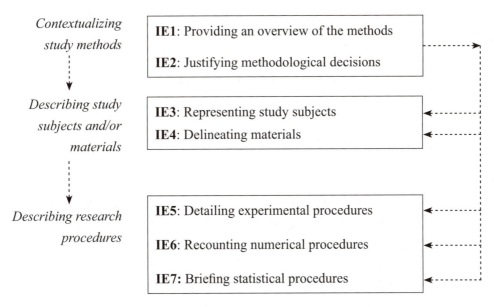

Figure 4.1   Information elements of the *Methods* section

Not all the IEs must appear in a paper. What elements are to be included is decided by how the study is designed. Some IEs, such as study subjects, materials and procedures, are inseparable by nature because research procedures are usually performed by using materials and/or study subjects. They are therefore frequently integrated and not clear-cut with each other.

> The *Methods* section is often composed of multiple subsections, whose headings may stand for specific procedures in most cases. These subheadings not only help writers organize this section but also assist readers in locating the information of interest.

Task 4.1 Read the *Methods* section of Sample Article 4 (SA4) again and answer the questions below.

1. What IEs do you identify? Which IEs are integrated with others?
2. Is it a good idea to omit Paragraph 2 in the subsection of *Study dataset*? And explain the possible reasons.
3. What tense and voice dominate this section? Why?
4. How do the authors organize the information in the subsection of *Process*?
5. Does Subsection 2.2 provide enough details for other researchers to duplicate the experiment?

Contextualizing study methods

> This information chunk, mostly located at the beginning of the *Methods* section, provides necessary background information for the research methods, thus setting the scene for describing study details and explaining procedural steps.
>
> It functions as a frame that encircles the description of the study and the analysis of the data.
>
> **IE1: Providing an overview of the methods**
> You may begin the *Methods* section by supplying some general information to substantiate the methodology used in the study. The information may encompass a summary of the *Methods* section, an overall announcement of the method and its purpose, a restatement of research purposes or hypotheses, gaps in empirical knowledge, and other information necessary to be introduced earlier before detailed descriptions and explanations of the methodology used. Below are some examples.

In this section, we first review the basics of Recurrent Neural Networks (RNNs) and the Long Short Term Memory (LSTM) model. Then we describe the specific algorithms for prediction of actions and forces used in our approach.

We developed a proof of concept (PoC) implementation of the decentralized access control system in order to test and evaluate it. The following section provides additional details about our implementation, in particular regarding the IoT devices, management hub, and the blockchain network.

IE2: Justifying methodological decisions

This information element aims to reason and explain various methodological choices or decisions including but not limited to the choice of subjects, materials, experimental actions, setting, etc. It may be present at the beginning of the *Methods* section to rationalize decisions made before the study or be embedded in the descriptions of other IEs to lend credibility to any choices made as the study unfolded. Below is an example.

In this section, we discuss the control design technique proposed to solve the multi-agent collision avoidance problem. Since Nash equilibria for the differential game introduced in Problem 2 cannot be easily obtained, a systematic method for constructing feedback control laws, which satisfy partial differential inequalities (PDIs) (instead of equations), leading to $\dot{\varepsilon}_a$ -Nash (instead of Nash) equilibria is provided.

Task 4.2 Here is the excerpt from the *Methods* section of Sample Article 8 (SA 8). Discuss with your partner how the authors contextualize the research methods and underline the words or expressions signalling your decisions.

A Numerical Investigation of Injection Pressure Effects on Wall-Impinging Ignition at Low-Temperature for Heavy-Duty Diesel Engine

2. Simulation methodology

1 The purpose of CFD combustion simulation is to represent and explain the phenomena that appeared in the experiments [17] to better understand the spray-wall combustion process low ambient temperature condition. **2** The target fuel is − 50# diesel and the physical properties of the target fuel, such as kinematic viscosity, density, and distillation curves, were published in the previous paper [18]. **3** The n-dodecane was used to replace the diesel for both the spray physical model and the chemical mechanism because of its similar properties to diesel. **4** This section gives a brief introduction of model selection, meshing, simulation conditions, and model validation.

2.1. Simulation setup

…

Task 4.3 Read the excerpt from the *Methods* section of Sample Article 12 (SA12) and answer the following questions.

Effects of GDI Injector Deposits on Sparay and Combustion Charactersistics Under Different Injection Conditions

2. Experimental setup and methods

1 The clean and fouled injectors used in this study are the same injector. **2** The PSA 1.2 L 3-cylinder GDI engine had been run for 55 h under indicated mean effective pressure (IMEP) of 3 bar and 2000 r/min to load deposits on the clean injector. **3** The injector of the second cylinder was disassembled for the flow rates test at the injection pressure of 10 MPa and a fuel temperature of 30 °C. **4** The injection pulse width ranges from 1 to 5 ms. **5** The injection was repeated for 1000 times for one measurement, and the total fuel mass was measured by a digital electronic scale. **6** The measurements were repeated five to eight times for each test point. **7** Fig. 1 shows the correlation of injector pulse width and fuel mass flow rate for the clean and fouled injector. **8** The reduction in the mass flow rate of the fouled injector changes with the pulse width. **9** The reduction in flow rate peaks at 2 ms with approximately 9.5% and gets the minimum 3.5% at 5 ms. 10 After obtaining the mass flow rate data, the pulse width can be adjusted appropriately in the engine test to ensure that both injectors inject the same mass of fuel.

2.1. Spray test system

…

2.2. Optical engine test systems

…

Questions:

1. What is this paragraph about?
2. Is it possible to delete this paragraph from the *Methods* section?
3. Is the tense appropriately used in Sentence 2?
4. What tense is used in Sentences 3 and 5? Why is this tense used?
5. What tense is used in Sentence 4 and Sentences 7–9? Why is this tense used?

Task 4.4 Read the following sentences and underline the parts explaining or reasoning the methodological decisions or choices. Circle the words or expressions signalling the justifications.

1. Although it was trained for the purpose of pulmonary tuberculosis detection, the model was verified by professional radiologists to be good enough to separate candidate patches for viral pneumonia.

2. Only the segmentation-related bounding box regression part was preserved, regardless of the classification results, because only the former task was required at this stage in this study.

3. Only the territory close to the middle of this cube contained maximum information about the focus of infection. Hence, only the center image together with its two neighbors of each cube was collected to represent this region for further classification steps.

4. However, the average length of straight channels is a redundant variable with respect to P1 since they are completely inversely proportional.

5. …where MEM1, MEM2 are electromagnetic torques of generators for diesel plant and micro-HPP; M12, M21 are the reciprocal torque due to the parallel operation of the first generator with respect to the second one and the second generator in relation to the first one.

6. In addition, the Sauter mean diameter (SMD) is also needed for the breakup model validation. Akop et al. [32] experimentally studied the fuel droplet characteristics when diesel fuel impinged on a flat wall and concluded that SMD is 9.82 μm at 40 MPa injection pressure with a nozzle of 0.17 mm. A larger nozzle diameter will increase SMD slightly, but the effect is very small [33]. Therefore, the SMD results obtained by Akop et al. [32] were used to validate the breakup model in this study.

7. However, under the condition of small and medium load, the introduction of EGR can improve the emission and even improve the power performance to some extent. Therefore,

the intake manifold absolute pressure (MAP) was kept at 43 kPa. Because blending hydrogen can significantly improve combustion and emission characteristics under lean burn condition, experiments were conducted with λ=1.0, 1.1, 1.2. To eliminate the discrepancy caused by other control parameters, the hydrogen injection pressure was 5 MPa and the hydrogen injection advance angle was 105 crank angle (CA) before top dead center (BTDC).

Describing study subjects and/or materials

The *Methods* section is traditionally named the "materials and methods section", which emphasizes the two aspects necessary to be addressed in an experiment. "Materials" refers to what was studied or used, while "Methods" means the specific procedures and tools used to collect and analyze data in a study. This may explain why describing study subjects and/or materials forms one of the information chunks included in the *Methods* section.

IE3: Representing study subjects

Study subjects refer to the entity or population under investigation. They could be human participants, animals or even inanimate objects observed, acted upon or subject to some treatment. To represent study subjects, such information is possibly provided as the source, the number, major characteristics, major functions, selection criteria, and other essential information. Below is an example.

Three types of disposable masks were tested: disposable medical masks (CHTC Jiahua Non-woven Co., Ltd., China), disposable surgical masks (from three locations: ESound Medical Device Co., Ltd., Anbang Medical Supplies Co., Ltd., and Yubei Medical Supplies Co., Ltd., China), and KN95-grade masks (the 3M 9502 and KF94 masks from the Republic of Korea). These masks are referred to herein as "JH," "YX," "AB," "YB," "3M," and "KF," respectively.

When human subjects are involved, it is routinely required to include a statement relating to informed consent and approval by the appropriate Institutional Review Board(s). When experimental animals are involved, a statement is needed to indicate that the experiments comply with the animal regulations of the institution concerned.

IE4: Delineating materials

The word "materials" is used as an umbrella term to represent natural or fabricated substances and tools assisting researchers in collecting and analyzing data to answer research questions. They may include chemicals, equipment, systems, algorithms, models, software, theorems, and other items utilized to facilitate the execution of the study.

To ensure potential replication and validity justification of a study, sufficient details about materials need to be provided, such as quantities, sizes, temperatures, treatment duration, specifications, conditions, locations, origin, chemical or physical properties, preparation, manufacturers, suppliers, and other key information. Below are some examples.

The path generation algorithm was realized using MATLAB. The library generation took about 13 days on an 8 GB-RAM computer with 2 CPUs (Intels Coret i5-8400, 2.8 GHz).

The filtration efficiency of the samples for sodium chloride (NaCl) particles was measured by using TSI 8130 equipment (TSI Incorporated, USA).

When the materials are specifically developed for the study at hand, their description is detailed in passages or even a separate subsection; otherwise, the mention of materials is usually embedded in the functional content of describing procedures. Besides descriptions, you are suggested to rationalize your choices of materials.

Task 4.5 Read the excerpts from the *Methods* sections of Sample Article 9 (SA9). Work in pairs and discuss the questions below the text.

Characteristics of Instantaneous Particle Number (PN) Emission from Hybrid Electric Vehicles Under the Real-World Driving Conditions

2. Materials & methods

2.1. Test vehicles

1 Four gasoline-electric hybrid light-duty passenger vehicles were tested in this paper, and all the vehicles are under good maintenance. **2** Standard China 6 commercial gasoline fuel, purchased from a certain supplier, is used for the tests. **3** The main characteristics of the test vehicles were shown in Table 1. **4** It should be pointed out that vehicle 1 is a prototype for China 6, with its original UCC replaced by the cGPF.

Questions:

1. What are the study subjects?
2. What sentence pattern is used to identify the subjects?
3. What information about the study subjects is provided besides the name?
4. Do you think the first sentence is concise? If not, please rewrite it.
5. Why did the authors illustrate that all the test vehicles are under good maintenance in the first sentence?
6. Do you think it is necessary to add specific information about the supplier in Sentence 2?
7. What is wrong with the third sentence?

Task 4.6 Read the excerpt from the *Methods* section of Sample Article 1 (SA1). Work in pairs and discuss the questions below the text.

Machine Learning-Assisted Design of Flow Fields for Redox Flow Batteries

2 Experimental

2.4 Experimental test

1 The tested VRFB system comprises a single cell, two external electrolyte tanks, pumps, and other accessories, and the configuration of the single cell is illustrated in Fig. S5 (ESI†). **2** HTS01, HTS04, and HTS02 flow fields and the two benchmarks (SFF and IFF) were fabricated by machining channels on the graphite plates. **3** The ion exchange membrane was DupontTM Nafions 212. **4** Both the negative and positive electrodes were 4 layers of Sigracets 39 AA carbon paper (the nominal thickness for each layer is about 280 mm), which had been thermally treated in a muffle furnace in ambient air at 400 1C for 10 hours. **5** During the assembly of the battery, the carbon paper electrodes were compressed to about 1000 mm.

6 The setup for the pressure drop test is illustrated in Fig. S6 (ESI†). **7** The water was used as the working medium and was circulated at various volumetric flow rates of 35–75 mL min^{-1} by a 2-channel peristaltic pump (LongerPump, BT100-1L). **8** A pressure transmitter

(Suzhou Xuansheng Meter Technology Co., Ltd, PCM950A) was used to measure the pressure difference between the inlet and outlet of the flow field plate.

9 The electrochemical test was carried out on a battery test instrument (Neware Technology Ltd). **10** When testing the charge–discharge performance, the cut-off voltage of charge and discharge processes was 1.65 V and 0.80 V, respectively. **11** Then the polarization curves were obtained starting from around 90% state of charge; the battery was discharged with increasing the current density and lasted 10 seconds at each test step until the voltage reached zero. **12** The volume of both the negative and positive electrolyte solutions was 20 mL. **13** The negative electrolyte solution contained 1.0 M V^{3+} and 3.0 M H_2SO_4; the positive electrolyte solution contained 1.0 M VO^{2+} and 3.0 M H_2SO_4. **14** Both electrolytes were circulated by a 2-channel peristaltic pump (LongerPump, BT100-1L). **15** In addition, nitrogen gas was used to exhaust the air in the battery setup to avoid undesirable side reactions.

Question:
1. What is the study subject?
2. What information about the study subject is provided besides the name?
3. What is Paragraph 2 about? What materials are used?
4. What is Paragraph 3 about? What materials are used?
5. What sentence patterns are used to mention the materials in Paragraphs 2 and 3?
6. Underline the sentences with parentheses. What information is listed in them? Is it a good idea to delete the information from these parentheses?
7. Which sentence is used to justify the material choice?

Describing research procedures

After describing the study subjects and/or materials, you may naturally move on to describe research procedures. What procedures are included in a study are attributed to its specific research design and contents. Even so, research procedures can be generally categorized into three types, experimental, numerical procedures, and statistical procedures according to the methods typically used in science and engineering research.

IE5: Detailing experimental procedures

Experimental procedures are related to physical experiments which are performed to gather data and address research questions or test hypotheses. A series of experimental actions are naturally carried out in a certain sequence. A detailed description of these actions and their sequence exactly illustrates how an experiment was completed in a step-by-step manner, thereby ensuring the understanding and potential duplication of the experiment by other researchers. Below are some examples.

In a typical procedure, boiling water was directly poured into the container at room temperature. The volume of water exceeded 80% of the total capacity of the container and the temperature was measured by a thermometer.

Vehicle 1, vehicle 3, and vehicle 4 were tested on the real-driving road, and vehicle 2 was tested on the chassis dynamometer using the speed/altitude spectrum recorded during real driving. Each vehicle was tested once. The tests were conducted on the normal workdays beyond rushing hours (around 8:00 am for vehicle 1, 2:00 pm for vehicle 2 and vehicle 3) by a professional driver.

IE6: Recounting numerical procedures

Besides experimental methods, numerical methods are also widely used in current research works. "Numerical methods" is considered a general term that includes all computational methods used in science and engineering fields. Therefore, various actions related to numerical calculations and their sequence need to be clearly and concisely recounted so that others can replicate or understand your experiment. Numerical procedures mainly present analysis and deduction of equations, and implementation/application of algorithms, involving extensive mathematical analyses. They may stand as separate (sub)sections under such headings as theoretical computation, mathematical modelling, numerical methods, computational methods, numerical simulation, and others. Below is an example.

The mathematical model describing the dynamics of changes in the parameters of generators in an energy system consisting of two machines (diesel generator and Micro-HPP) can be described by the following system of equations [31]:....

> **IE7: Briefing statistical procedures**
>
> Statistical methods are used in a few studies. How to conduct a statistical analysis by using statistical software or techniques should be identified without mentioning the details of the calculative process. Below is an example.
>
> *Independent-sample t-tests were adopted to determine whether the difference between two grouped samples was significant or not. Linear regressions were performed to explore the positive or negative impacts of various explanatory variables (Section 3.4) on the spatiotemporal variations in OC transport. ... Statistical significance was considered at p < 0.05 (two-tailed test).*
>
> No matter what procedures you describe, supplying sufficient details helps to establish the credibility of your study. A new or unique method deserves elaboration with full details and justifications provided where necessary. A conventional or standard method, by contrast, will be briefly introduced by identifying its name and listing a literature reference. However, if any modifications have been made, a detailed description is also needed.
>
> As usual, most papers include multiple chronologically-described research steps. These steps can be grouped by the relevance existing among them and presented in separate subsections. The subheadings normally stand for the key procedures carried out to obtain results. In each subsection, the specific steps are still chronologically described.

Task 4.7 Read the excerpt from the *Methods* section of Sample Article 5 (SA5). Work in pairs and discuss the questions below the text.

Can Masks Be Reused After Hot Water Decontamination During the COVID-19 Pandemic?

2. Materials and methods

2.2. Hot water decontamination and charge regeneration

1 Three kinds of containers, including a household aluminum basin, a polypropylene plastic lunch box, and a stainless steel thermos cup, were used in the experiments. **2** In a typical procedure, boiling water was directly poured into the container at room temperature. **3** The volume of water exceeded 80% of the total capacity of the container and the temperature was measured by a thermometer. **4** The mask was immersed in the water by placing a heavier object on top of it, such as a spoon. **5** The container was then closed and the mask was left to soak in the hot water for 30 min. **6** After that, the container was opened and the mask was removed from the hot water. **7** The liquid on the mask was slightly shaken off and the mask was placed on the surface of dry insulating material, such as wooden, plastic tables, and bed sheets. **8** The mask was then dried with a standard hair dryer for 10 min.

2.3. Static electricity test

9 A hand-held electrostatic field meter (FMX-004; Simco, Japan) was used to test the electrostatic charge of mask. **10** The mask to be tested was hung on an insulation component at least 5 m away from other instruments with static electricity in order to avoid interference from other static electricity fields. **11** Before measurement, the researcher washed his or her hands with water to remove static electricity from the hands. **12** The probe was gradually moved closer to the measurement position on the mask until the two laser dots from the electrostatic field meter coincided. **13** The value of the instrument readings was recorded.

2.4. Waterproof test

14 Waterproof testing of the masks was performed by a Buchner funnel procedure. **15** To summarize, the mask was placed at the bottom of the bowl, with the outer surface of the mask facing upward. **16** A hose pipe was used to attach the filter flask sidearm onto the vacuum aspirator. **17** A vacuum aspirator operating at 30 $L \cdot min^{-1}$ was used for the suction of liquids, through the filter paper. **18** A total of 100 mL water was then poured onto the surface of mask within 20 s from a height of 20 cm. **19** The pumping filtration was maintained for 3 min and the flask was watched to see if water dropped down. **20** Then the water was removed from the Buchner funnel, and the vacuum suction system was turned off. **21** The inner surface of the mask was then observed to determine the wettability/waterproofness of the mask.

2.5. Filterability test

22 The filtration efficiency of the samples for sodium chloride (NaCl) particles was measured by using TSI 8130 equipment (TSI Incorporated, USA). **23** The particle size distribution of the NaCl aerosol for the specified test conditions was the median diameter of the number of particles at (0.075 ± 0.020) lm. **24** The geometric standard deviation did not exceed 1.86 and the concentration did not exceed 200 mg · m^{-3}. **25** The detection system had a device to neutralize the charged particles. **26** To test the KN95-grade masks, the gas flow rate was stabilized to (85 ± 4) L · min^{-1}, while to test the disposable medical masks and surgical masks, the gas flow rate was stabilized to (30 ± 2) L · min^{-1}, and the cross-sectional area of the air flow was 100 cm^2.

Questions:
1. What is the relationship between these subsections? Can their order be changed and why?
2. What temporal words or expressions are used in these sections? Why are they used?
3. Highlight the words or expressions signalling the materials used and the actions performed in Subsection 2.2. What could you learn from them about how to describe materials and procedures?
4. Why are the underlined words or expressions used in Subsection 2.2?
5. What IEs are included in Subsection 2.3?
6. Which sentences are used to justify the methodological decisions in Subsections 2.4 and 2.5?
7. How are the sentences organized in Subsection 2.4? Does it hold true for other subsections?

Task 4.8 Read the excerpt from the *Methods* section of Sample Article 1 (SA1). Work in pairs and discuss how the authors describe the numerical procedure compared with the experimental procedures described in Task 4.7.

Machine Learning-Assisted Design of Flow Fields for Redox Flow Batteries

2 Experimental

2.1 Path generation algorithm

1 A path generation algorithm was developed to generate a library containing as many as 11 564 flow field designs, represented by 2-D binary images. **2** This algorithm starts with an undirected graph with a rectangular grid of cells and continuously extends a path between the inlet and the outlet until a Hamiltonian path is found. **3** Specifically, the initialization of the generation process is shown as follows: (I) given the inputs, the number of rows and columns of the grid, n_1 and n_2, and the cell size, δ, are calculated. **4** (II) An undirected graph G with an $n_1 \times n_2$ rectangular grid of cells is generated (Fig. 2(a1)). **5** For each cell, a vector is used to record if the surrounding walls exist and if the cell is already visited. **6** (III) An initial path G^0 (e.g., diagonal), which connects the inlet with the outlet, is generated (Fig. 2(a2)). **7** If two adjacent cells are connected, the wall between them will be removed, and meanwhile, the edge will be added accordingly. **8** After the initialization, the program enters the main loop. **9** Assuming $k-1$ iterations have been completed, the evolution of the current path G^k includes the following steps: (IV) all the edges of the current path, $e_1,...,e_p$, are identified, and then the possible moving directions for each edge, Γ, are determined (Fig. 2(a3)). **10** For a horizontal edge, ... **11** Note that a moving direction will be considered infeasible if it makes the edge exceed the entire grid or encounter the existing path. **12** (V) A candidate path $G^k|_{ei,\Gamma}$ is obtained by moving ei of G^k along Γ (Fig. 2(a4)). **13** Then the energy of the candidate path, $E(G^k|_{ei,\Gamma})$, is computed. **14** Here E is a hypothetical energy operator, which is given by:

$$E(G^k|_{ei}, \Gamma) = \frac{N_{turn}}{(L_{path})^2} \quad (1)$$

where N_{turn} is the number of turns of the path, and L_{path} is the length of the path. ...

Task 4.9 Rewrite the following short sentences describing procedures into long sentences.

1. The two laser dots from the electrostatic field meter coincided. Then the probe was gradually moved closer to the measurement position on the mask.

2. Both the negative and positive electrodes were 4 layers of Sigracets 39 AA carbon paper (the nominal thickness for each layer is about 280 mm). The paper had been thermally treated in a muffle furnace in ambient air at 400 1C for 10 hours.

3. To test the KN95-grade masks, the gas flow rate was stabilized to (85 ± 4) L · min^{-1}. To test the disposable medical masks and surgical masks, the gas flow rate was stabilized to (30 ± 2) L · min^{-1}, and the cross-sectional area of the air flow was 100 cm^2.

4. Fig. S1a (ESI†) illustrates the uniformity factor. The uniformity factor quantifies the uniformity of the distribution of electroactive species within the porous electrode. The uniformity factor was calculated by equation(11).

$$y_1 = 1 - \frac{1}{c_{i,avg}} \sqrt{\frac{1}{V_e} \iiint (c_i - c_{i,avg})^2 dV_e} \quad (11)$$

5. Equation (9) describes the charge transport in the electrode. Equation (10) describes the electrolyte.

$$\nabla \cdot \vec{t}_s = -\sigma_s^{eff} \nabla^2 \phi_s = i_{loc} \qquad (9)$$

$$\nabla \cdot \vec{t}_1 = -\sigma_1^{eff} \nabla^2 \phi_1 = -i_{loc} \qquad (10)$$

$\vec{t}_s$ is the local current density in the electrode. $\vec{t}_1$ is the local current density in the electrolyte. σ_s^{eff} is the effective electronic conductivity of the electrode. σ_1^{eff} is the effective electronic conductivity of the electrolyte.

LANGUAGE CONVENTION

Form groups of four. Each member picks one sample article listed in the table below and reads the *Methods* section. Fill in your part of the table and compare your answers within the group.

Article	Tense	Voice	Frequently used sentence pattern
SA1			
SA2			
SA4			

Verb tenses

> The *Methods* sections are mostly written in the simple past tense/simple present tense. The choice of verbal tenses will depend on your discipline and the actions you are describing.

Unit 4 Describing Your Methods

The simple past tense is generally used to report research actions, describe materials, and especially to illustrate the procedures, because the methods and materials used, procedures followed and data analysis are all enacted at a particular time and condition. However, the simple present tense is often used to describe standard or conventional materials or procedures, existing datasets or samples.

If you are still in doubt, you can refer to the recently published articles in your target journals to get an idea of what tense is dominantly used in the *Methods* section.

Task 4.10 The following is an excerpt from the *Methods* section of a research paper in the field of environmental engineering. Fill in the blanks with the appropriate tenses and voices of the given verbs.

1 The current investigation involved sampling and analyzing six sites to measure changes in groundwater chemistry. **2** The sites _____ (select) from the London Basin area, which _____ (be) located in the southeast of England and has been frequently used to interpret groundwater evolution.

3 A total of 18 samples _____ (collect) and then _____ (analyze) for the isotopes mentioned earlier. **4** Samples 1–9 _____ (collect) in thoroughly-rinsed 25 ml brown glass bottles which _____ (fill) to the top and then sealed tightly to prevent contamination. **5** The filled bottles _____ (ship) directly to two separate laboratories at Reading University, where they _____ (analyse) using standard methods suitably miniaturised to handle small quantities of water. [5]

6 Samples 10–18 _____ (prepare) in our laboratory using a revised version of the precipitation method established by the ISF Institute in Germany. **7** This method _____ (obtain) a precipitate through the addition of $BaCl_2 \cdot 2H_2O$; the resulting precipitate _____ (can) be washed and stored easily. **8** The samples subsequently _____ (ship) to ISF for analysis by accelerator mass spectrometry (AMS). **9** All tubing used _____ (be) stainless steel, and although two samples were at risk of CFC contamination as a result of brief contact with plastic, variation among samples _____ (be) negligible.

Task 4.11 Read the following excerpts from Sample Articles. Underline the verbs used in each sentence and discuss with a partner why a specific tense was chosen.

1. A path generation algorithm was developed to generate a library containing as many as 11 564 flow field designs, represented by 2-D binary images. This algorithm starts with an undirected graph with a rectangular grid of cells and continuously extends a path between the inlet and the outlet until a Hamiltonian path is found.

2. As illustrated in Fig. S1a, the uniformity factor, which quantifies the uniformity of the distribution of electroactive species within the porous electrode, was calculated by:

$$y_1 = 1 - \frac{1}{c_{i,\text{avg}}} \sqrt{\frac{1}{V_e} \iiint (c_i - c_{i,\text{avg}})^2 \, dV_e} \tag{11}$$

3. As shown in Fig. 2 and Fig. 3, when the KH breakup time constant is 50 and the RT breakup length constant is 25, the spray penetration, impinged radius and the SMD are all matched well.

4. According to the study of the flame spectra [36], the incandescence of the soot dominates the luminance over 600 nm. Therefore, a red filter (long pass filter) with a cutoff wavelength of 600 nm was used to highlight the incandescence of the soot.

5. Six datasets of catchment characteristics (①–⑥) were obtained to quantitatively explain the spatiotemporal variations in riverine OC transport. ①–③ are the monthly mean land surface temperature (LST,°C), monthly mean vegetation coverage indicated by NDVI, and annual land cover during 2004–2018, respectively. They were extracted from the remote sensing datasets of MOD11A1, MOD13A2, and MCD12Q1, produced using MODIS/Terra data (https://lpdaac.usgs.gov/).

Passive structures

Passive structures play an important role in scientific writing, especially in the *Methods* section because they shift emphasis from actors to actions, thus making the writing impersonal. Nevertheless, such structures are more wordy and less readable than active structures. Their overuse makes writing difficult to understand. Therefore, you are suggested to use active and passive structures appropriately in your papers.

However, it is not simple to make a proper choice between active and passive structures. Apart from the preference and trend of scientific writing, the following three questions are worth your consideration in your choice.

1. What information needs to be emphasized?
Different information can be emphasized by being set as the grammatical subject of the verb. When writing a sentence, you need to decide which part to emphasize, which contributes to the choice of active or passive voice. See the examples below.

To emphasize the actor	<u>We</u> performed the 3-D multi-physics simulation on the software COMSOL Multiphysics.
To emphasize the action	The 3-D multi-physics <u>simulation</u> was performed on the software COMSOL Multiphysics. (SA1)
To emphasize the tool	The <u>software</u> COMSOL Multiphysics was used to perform the 3-D multi-physics simulation.

2. Is the sentence easy to understand?
A common problem with writing passive sentences is top-heavy sentences. Top-heavy sentences are passive sentences with a too long subject, but a short passive verb right at the end. These sentences are usually uneasy to read. The solution to this problem is getting the subject and the verb within the first few words of the sentence and placing any list of items at the end. Compare the two sentences below.

Top-heavy	*Wheat and barley, collected from the Virginia field site, as well as sorghum and millet, collected at Loxton, were used.*
Improved	*Four types of cereals were used: wheat and barley, collected from the Virginia field site; and sorghum and millet, collected at Loxton.*

3. Does the choice help the inter-sentence connection?

Effective writers tend to employ various strategies to connect the sentences, thus achieving textual coherence. One strategy is putting old information near the beginning of the sentence and new information at the end. Old information means something mentioned or related to an idea mentioned in the preceding text. Sometimes the choice of a passive sentence is due to the old-and-new-information arrangement. See the examples below.

Original	*A 20-item version of Big-Five Inventory (BFI-20), which was developed by Engvik and Clausen (2011), was used to measure <u>personality traits. They</u> constructed each trait on the basis of several items.*
Improved	*A 20-item version of Big-Five Inventory (BFI-20) developed by Engvik and Clausen (2011) was used to measure <u>personality traits. Each trait</u> was constructed on the basis of several items.*
Improved	*Personality traits were measured by a 20-item version of Big-Five Inventory (BFI-20) developed by <u>Engvik and Clausen (2011). They</u> constructed each trait on the basis of several items.*

Task 4.12 Identify what information is emphasized in the following sentences. Please work out a different version for each sentence by shifting the sentence focus.

1. The test set was used to evaluate the performance of the trained model.

   Emphasis from _____ to _____

   Your sentence: _____

2. The CNN described above was developed using the deep learning toolbox from MATLAB.

   Emphasis from _____ to _____

   Your sentence: _____

3. A 3-D multi-physics model was used to calculate the uniformity factor and pressure drop of 1164 flow fields in the library.

   Emphasis from _____ to _____

   Your sentence: _____

4. We recorded the R-squared R2 and the mean absolute percentage error (MAPE) for the test set to evaluate the prediction performance of the CNNs.

   Emphasis from _____ to _____

   Your sentence: _____

5. The individual weights were then combined using geometric and arithmetic means.

   Emphasis from _____ to _____

   Your sentence: _____

6. The overall analysis report for one CT sample was calculated using the Noisy-OR Bayesian function [25].

   Emphasis from _____ to _____

   Your sentence: _____

Task 4.13 Improve the following top-heavy sentences to make them easier to read.

1. Simulation results at the virtual furnace exit and reference data provided by Shangdu power plant were compared.

2. A balance between deep and shallow rooting plants, heavy and light feeders, nitrogen fixers and consumers and an undisturbed phase is needed to achieve maximum benefit through rotation.

3. Finally, the set of criteria related to geographical, economic, ecological, and climatic conditions was selected as follows.

4. National Cancer Institute Common Terminology Criteria for Adverse Events version 4.03 was used to grade adverse events (AEs).

5. The five traits, their definitions according to American Psychology Association (APA) dictionary of psychology (American Psychological Association (APA), 2007), and the items associated with each trait are shown in Table 1.

6. The demographic characteristics (gender, age, years of education, height, body mass, body mass index, PA level, and the amount of time spent sitting) among the high, moderate, and low PA groups were compared by using a one-way analysis of variance (ANOVA).

7. Actual evapotranspiration (T) for each crop, defined as the amount of precipitation for the period between sowing and harvesting the particular crop plus or minus the change in soil water storage in the 2 m soil profile, was computed by the soil water balance equation (Xin, 1986; Zhu and Niu, 1987).

Task 4.14 Read the following excerpts from Sample Article 1 (SA1). Discuss with a partner how passive or active structures are used for inter-sentence connection.

1. As mentioned in the path generation algorithm, the flow field is represented by a rectangular grid of cells, and these cells fall into two categories:[60] "straight" and "turn" (Fig. S7). The turn cells divide the flow channel into several sections; these sections are henceforth called straight channels.

2. The negative electrolyte solution contained 1.0 M V^{3+} and 3.0 M H_2SO_4; the positive electrolyte solution contained 1.0 M VO^{2+} and 3.0 M H_2SO_4. Both electrolytes were circulated by a 2-channel peristaltic pump (LongerPump, BT100-1L).

Characteristic expressions

The characteristic expressions frequently used in the *Methods* section can be categorized to perform the following functions:
- Providing an overview of the methods;
- Describing materials;
- Describing the research procedures;
- Justifying methodological decisions;
- Recounting statistical or numerical procedures.

Scan the QR code for a list of the expressions. Try to get familiar with them and pick some to use in your future writing.

Task 4.15 Translate the following sentences into English.

1. 我们使用多重通用线性模型（GLM）定量计算了不同因素对河流有机碳（riverine organic carbon）输送变化的相对贡献。

2. 训练卷积神经网络（CNN）来捕获输入变量到输出变量的函数映射（functional mapping）。

3. 为了分析流场（flow field）的设计规则，我们通过定义五种几何特性来描述流道（flow

channel）的形态特征。几何特性运用矩阵实验室（MATLAB）进行计算。

4. 本实验采用独立样本 t 检验来确定两个分组样本之间的差异是否显著。$P<0.05$ 时，差异具有统计学意义。

5. 这项研究征得该大学伦理委员会同意，并遵守了相关指南和规定。在研究开始前，所有参与者和 / 或其法定监护人签署了知情同意书。

Check your understanding

Task 4.16 Choose two from the Sample Articles (SA1–SA12) and read the *Methods* sections. Fish the tasks below.

1. Analyze the IEs included.
2. Locate the IEs not presented in detail and try to explain the possible reasons.
3. Analyze the tenses and voices used.
4. Highlight characteristic expressions.

Task 4.17 Pick two research papers in your field. Analyze the *Methods* sections and finish the tasks below.

1. Read through the *Abstract* of the chosen papers to have a general understanding of the *Methods* section.

2. Read the *Methods* section carefully by
   - relating sub-headings to the information chunks;
   - locating the subjects and/or materials used;
   - highlighting the criteria for or restrictions on the selection of the subjects and/or materials if applicable;
   - specifying the research procedures and deciding how they are logically arranged;
   - identifying statistical or numerical analysis, tools, and/or procedures if applicable;
   - examining whether each IE is presented in detail or not;
   - noting down characteristic expressions for each IE.

3. Summarize your analysis orally or in writing for class discussion next time.

Unit task

Drafting Your *Methods* section

Up to now, you have already known about the general function of the *Methods* section, understood the common information elements in this section and learned about the linguistic strategies for writing an effective *Methods* section. It is time to set about drafting the *Methods* section for your own paper. Do the following to finish this task.

Step 1: Check the *Introduction* section you have written for your research paper to remind yourself of the research purpose.

Step 2: Select from your reference list the one highly relevant to your study. Read the *Methods* section carefully and analyze the information elements included in it. Underline the characteristic expressions for each information element.

Step 3: Take the *Methods* section of the chosen research paper as a template and write your own *Methods* section.

Step 4: Check the *Methods* section you have written by answering the following questions:
- Could the methods be duplicated by readers to produce the same results?
- Did I use appropriate materials?
- Did I justify my own methods and actions?
- Is the purpose told clearly when a particular procedure is introduced?
- Are tenses used appropriately?

Unit 5

Presenting Your Results

Learning objectives

In this unit, you will
- understand the general function and purposes of the *Results* section;
- learn about the common information elements in the *Results* section;
- develop the linguistic strategies for writing an effective *Results* section.

Self-evaluation

Read through the *Results* section of Sample Article 2 (SA2) "The Site Selection of Wind Energy Power Plant Using GIS-Multi-Criteria Evaluation from Economic Perspectives" and answer the following questions.
- Does this section provide the information you expect?
- How do the authors organize this section?
- Can you organize the section differently but more effectively?
- Why are the graphics used? Would it be good to use text instead?
- How are the graphics referred to in this section?

The *Results* section is essential in expressing the significance of a research project. It is the most effective part that makes the research distinctive and also the most attractive part to editors, reviewers, and readers. The function of this section is to accurately and objectively present the observed or measured results of the research. The information presented here is very specific. In many cases, this section takes the most pages in the paper.

To be clear and effective, authors often employ both words and visuals to illustrate their results. Therefore, the *Results* section is featured by graphics such as tables, figures, charts, graphs, diagrams, images, and photographs. The graphics are mentioned in the text and the major information presented in the graphics is highlighted. Sometimes, an introductory paragraph or some sentences may appear at the beginning of the section, briefly restating the research aim, research question, or principal activities in the research. Such a component serves as a transition from the preceding *Methods* section to the current section.

Presenting study results is of great importance in a research report. Writing an effective *Results* section is thus crucial to a paper. This unit will help you understand and grasp how to write an acceptable *Results* section in terms of structure and language.

INFORMATION CONVENTION

Below is the *Results* section of Sample Article 4 (SA4). Read it and answer the following questions.

A Deep Learning System to Screen Novel Coronavirus Disease 2019 Pneumonia

4. Results

4.1. Evaluation platform

1 An Intel i7-8700k central processing unit (CPU) with NVIDIA graphics processing unit (GPU) GeForce GTX 1080ti was used as the testing server. **2** The processing time largely depended on the number of image layers in one CT set. **3** On average, it took less than 30 s for a CT set with 70 layers to go from data preprocessing to the output of the report.

4.2. Training process

4 As one of the most classical loss functions used in classification models, cross entropy was used in this study. **5** When the epoch number of training iterations increased to more than 1000, the loss value did not obviously decrease or increase, suggesting that the models converged well to a relative optimal state without distinct overfitting. **6** The training curves of the loss value and the accuracy rate for two classification models are shown in Fig. 5. **7** The model with the location-attention mechanism achieved better performance on the training dataset, in comparison with the original ResNet.

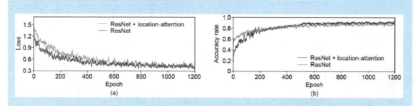

Fig. 5. Training curve of (a) loss and (b) accuracy rate for the two classification models.

4.3. Performance on test dataset
4.3.1. Performance measurement

8 A confusion matrix was used, which is a table that is often used to describe the performance of a classification model on test dataset for which the true values are known. **9** It allows the visualization of the performance of an algorithm.

10 The accuracy (A) of a method determines how correct the predicted values are. **11** Precision (P) determines the reproducibility of the measurement, or how many of the predictions are correct. **12** Recall (R) indicates how many of the correct results are discovered. **13** The f_1-score uses a combination of precision and recall to calculate a balanced average result. **14** The following equations show how to calculate these values, where TP, TN, FP, and FN are true positive, true negative, false positive, and false negative, respectively.

$$A = \frac{TP+TN}{TP+FP+TN+FN} \tag{3}$$

$$P = \frac{TP}{TP+FP} \tag{4}$$

$$R = \frac{TP}{TP+FN} \tag{5}$$

$$f_1\text{-score} = \frac{2 \times P \times R}{P+R} \tag{6}$$

4.3.2. Image preprocessing and segmentation

15 A total of 90 CT samples were randomly selected from each group (30 CT sets from COVID-19, 30 from IAVP, and 30 from healthy cases) for the test dataset. **16** The choice of the test dataset followed the rule that any CT of this person had not been trained in the previous stage, in order to avoid having a similar CT that had been learned by the models. **17** Moreover, the thresholds for both the image preprocessing and the segmentation were optimized to be more suitable for the current study. **18** In the image preprocessing stage, the threshold of the Hounsfield unit (HU) value, which was used to binarize the resampled images, was raised to -200 in order to maximize the filtering out of valid lung. **19** The segmentation model VNet–IR–RPN was configured to reduce the proposal's threshold to maximize separate candidate regions, even through many normal regions could be included. **20** We noticed that one CT case from the COVID-19 group that had no image patches was segmented as COVID-19 or IAVP, and was hence wrongly categorized as healthy case, as shown in Fig. 6. **21** These focuses of infection were barely noticeable with the human eye, and seemed too tenuous to be captured by the segmentation model in this study.

binarize 二值化

image patch 图像块

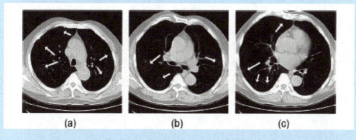

Fig. 6. All CT images (a-c) were from a single CT case. The focuses of infections were pointed out by arrows.

4.3.3. Classification for a single image patch

22 A total of 1710 image patches were acquired from 90 CT samples, including 357 COVID-19, 390 IAVP, and 963 ITI (ground truth). **23** To determine the optimal approach, the design of each methodology was assessed using a confusion matrix. **24** Two classification models were evaluated: with and without the location-attention mechanism, as shown in Tables 1 and 2.

Table 1
Confusion matrix of two classification models for the COVID-19, IAVP, and ITI groups.

Group	COVID-19 (M_1/M_2)	IAVP (M_1/M_2)	ITI (M_1/M_2)
COVID-19 (M_1/M_2)	260/273	47/32	50/52
IAVP (M_1/M_2)	55/46	276/280	59/64
ITI (M_1/M_2)	75/77	81/82	807/804

M_1: the ResNet model; M_2: the ResNet model with the location–attention mechanism.

Table 2
Recall, precision, f_1-score, and accuracy rate[a] of two classification models for the COVID-19, IAVP, and ITI groups.

Group	R	P	f_1-score
COVID-19 (M_1/M_2)	0.728/0.765	0.667/0.689	0.696/0.725
IAVP (M_1/M_2)	0.708/0.718	0.683/0.711	0.695/0.714
ITI (M_1/M_2)	0.838/0.835	0.881/0.874	0.859/0.854

[a] The overall accuracy rates of M_1 and M_2 for the three groups are 78.5% and 79.4%; the accuracy rates of M_1 and M_2 for COVID-19 and IAVP groups are 71.8% and 74.0%.

25 The average f_1-score and the overall accuracy rate for the two models were 0.750/0.764 and 78.5%/79.4%, respectively. **26** Furthermore, the location-attention mechanism was used to improve the respective accuracy rate of the COVID-19 and IAVP groups, and was shown to result in a remarkable improvement of 5.0% (260/273) and 1.4% (276/280). **27** This evidence indicates that the second model with the location-attention mechanism achieved better performance. **28** Therefore, that model was used for the rest of this study.

29 Moreover, as the ITI group was used to remove disturbing factors in this study, it was ignored and not counted by the Noisy-OR Bayesian function in the final step. **30** To retain consistency in the next steps, we further compared the average f_1-score and accuracy rate for the first two groups, which were 0.720 and 74.0%, respectively.

4.3.4. "Voting" for a region

31 Each image patch "voted" to represent the entire candidate region. **32** A total of 570 candidate cubes were distinguished, including 119 COVID-19, 130 IAVP, and 321 ITI regions (ground truth). **33** The confusion matrix of the voting result and the corresponding recall, precision, and f_1-score are shown in Tables 3 and 4, respectively.

34 The average f_1-score and overall accuracy rate for the three categories were 0.856 and 89.3%, respectively, which showed a respective improvement of 12.0% and 12.5% when compared with the previous step. **35** As for the first two groups, the average f_1-score and accuracy rate were 0.806 and 78.3%, respectively, which showed a respective increase of 11.9% and 5.8%.

4.3.5. Result of the classification of CT samples as a whole

36 Noisy-OR Bayesian function was used to identify the dominating infection types. **37** Three kinds of results were exported in the final report: COVID-19, IAVP, and healthy cases. **38** The experimental results are summarized in Tables 5 and 6.

39 Only the average f_1-score and the accuracy rate of the COVID-19 and IAVP groups were counted to compare with the previous results. **40** These were 0.843 and 85.0%, respectively, which showed a promotion of 4.6% and 8.6% in this step.

41 A consistent improvement of the average f_1-score and accuracy rate was observed. **42** The accuracy rate of the classification of COVID-19 and IAVP was promoted from 74.0% (single image patch) to 78.3% (image cube), and then to 85.0% (overall CT case based on the dominating infection types). **43** Measured by all three benchmark groups, the overall classification accuracy rate was 86.7%.

44 Moreover, a series of images with highlighted focuses of infection would also be exported (examples shown in Fig. 7).

Questions:
1. How do the authors present the information in this *Results* section?
2. Why do the authors use tables and figures?
3. How are the tables and figures numbered?
4. What is the function of the note under a table or figure? Is the information in the note repeated in the text?
5. Can you find some sentences where tables and figures are mentioned? How are the tables and figures referred to?
6. Can you find some sentences reporting the study results? Does the text describe whatever is shown in a table or figure?
7. Are there any sentences used to comment on the results?

Overall structure of the *Results* section

Though what is covered in the *Results* section may vary with disciplines and journals, some basic information components are shared by most research papers. The information elements (IEs) that are conventionally included in the *Results* section can be classified into three information chunks: indicating the research procedure(s) where the result(s) under discussion was obtained, presenting the result(s), and commenting on the results. The three chunks often appear in cycles.

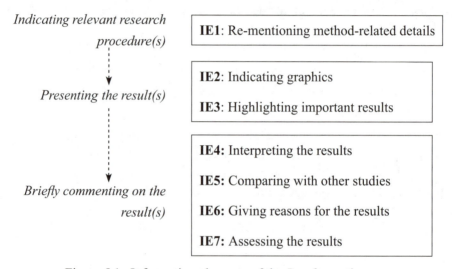

Figure 5.1 Information elements of the *Results* section

Not all the IEs may appear in a paper. IE2 and IE3 are essential in the *Results* section. The information chunk, briefly commenting on the results, is more likely to appear in the *Discussion* section. Only in a few cases will authors make some comments on the findings in the *Results* section. Therefore, this unit will focus on the first and the second information chunks.

Task 5.1 Read the *Results* section in Sample Article 4 (SA4) again and analyze what information element(s) each sentence indicates. Complete the table below.

Sentence No.	IE	Sentence No.	IE	Sentence No.	IE
1		16		31	
2		17		32	
3		18		33	
4		19		34	
5		20		35	
6		21		36	
7		22		37	
8		23		38	
9		24		39	
10		25		40	
11		26		41	
12		27		42	
13		28		43	
14		29		44	
15		30			

Indicating relevant research procedures

Research projects vary in their purposes, steps, and findings. In most papers, multiple results are presented, such as major findings and minor ones, initial outcomes and incremental ones, or individual results and conclusive ones. When reporting specific results, authors should make it clear how and/or where the results were obtained. This can be realized by reminding readers of the details of the relevant procedures. This element is particularly needed when the research procedures are complicated and multiple results are to be presented.

IE1: Re-mentioning method-related details

The mention of method-related details helps readers recall the principal research activity and better understand through what procedures a specific aim was achieved and the corresponding results were obtained. In many cases, justifications of the methods are provided, which adds credibility to the obtained results. Two important ways to justify the methods are the use of infinitive phrases and the expression like "the reason for choosing this method is that…." Below are some examples.

To generate a custom library of flow field designs, we developed a path generation algorithm (see Section 2.1).

Furthermore, to analyze the working mechanism of the well-trained CNNs, we visualized the networks by plotting the feature maps of the convolutional and max-pooling layers, as shown in Fig. 3(e) and (f).

A CDCS strategy is used to distribute power between the battery and APUs.

The reason for choosing this method is that it can address…

Task 5.2 Read the *Results* section of Sample Article 2 (SA2) "The Site Selection of Wind Energy Power Plant Using GIS-Multi-Criteria Evaluation from Economic Perspectives." Identify the method-related details in each subsection and underline the words or expressions used to justify the methods. Complete the table below.

Paragraph	Method-Related Details
1	
2	
3	
4	

Task 5.3 Answer the questions below based on the method-related details you have identified in Task 5.2.

1. Would it be a good idea to skip these details and report the results directly? Why?
2. Skim the *Methods* section again to check whether the method-related details mentioned in the *Results* section have been reported or not. If not, please use one specific example to illustrate why sometimes method-related details are presented in the *Results* section.

Presenting the results

> Presenting the results is essential to this section. Authors often present results with both words and graphics. Visualization of results has become increasingly important. Effective graphics could be decisive to the "fate" of a paper. Equally important is the corresponding text that describes the results and explains what is shown by the graphics. In brief, two major informational elements are included to fulfill the task of presenting study results.

IE2: Indicating graphics

Although graphics are effective and visually impactful in illustrating results, it is still necessary to describe and explain the results with words. When doing so, you need to refer to the graphics in your text. There are different ways to indicate graphics. You can use a complete sentence or a dependent clause to indicate in which table or figure the results are shown or can be found. You can also place the figure number in parentheses after the statement of results. Remember, the simple present tense should be used when you indicate a graphic with a complete sentence because the sentence is telling a fact. Below are some examples.

The confusion matrix of the voting result and the corresponding recall, precision, and f_1-score are shown <u>in Tables 3 and 4</u>, respectively.

<u>Table 5 lists</u> the optimal purchasing price for small and large power plants with respect to the geographical location of cities.

Two classification models were evaluated: with and without the location attention mechanism, <u>as shown in Tables 1 and 2</u>.

<u>As Fig. 9 shows</u>, the idle mode experiment demonstrates how a turbine with electromagnetic CVT picks up speed.

For the Yangtze River, the POC concentrations at different sites decreased significantly ($p < 0.05$), but the COD/POC ratios increased apparently ($p < 0.05$) during 2004-2018 <u>(Fig. 5)</u>.

The sentences indicating graphics can be further classified into two forms: indicative statements and informative statements. An indicative statement summarizes what kind of research was done, while an informative statement highlights specific research findings. The former serves as a start of the reporting, followed by other sentences giving detailed results. The latter, however, moves faster to present specific details of the result directly. A common way to introduce an informative statement is using the "as-clause." See the examples below.

Indicative statement	*Table 4 shows the <u>changes</u> in the number of buffalo in the Serengeti National Park since 1965.*
Informative statement	*Table 4 shows that the number of buffalo in the Serengeti National Park <u>decreased</u> sharply since 1965.*
Informative statement	*<u>As shown in Table 4</u>, the number of buffalo in the Serengeti National Park decreased sharply since 1965.*
Informative statement	*<u>As can be seen from Figure 3</u>, gender did not account for any differences in perceived organizational culture and commitment.*

IE3: Highlighting important results

Highlighting important results is the dominant component of the *Results* section. The authors focus on presenting the specific values, changes, fluctuations, and trends of variables, as well as the similarities or differences and the relationships between or among variables. Therefore, the occurrence of numbers, symbols, units, equations, etc. is very common in the *Results* section.

Highlighting important results is like telling a story. It goes paragraph by paragraph and follows a logical order. The findings of the research may be presented in the order of prominence or the order in which the experimental steps were carried out. It is unnecessary to report every detail shown in graphics; only the most salient information needs to be highlighted. A common practice is to organize the information into subsections. This is helpful because readers can acquire a high-level summary of the results by just reading the subheads (e.g. SA3).

Task 5.4 Read the following sentences and identify different expressions indicating graphics. Pay attention to the verbs, tense, and voices in these expressions. Rewrite Sentence 2 using different expressions indicating the graphic and complete the table below.

1. Fig. 5(a) shows the charge–discharge curves of the VRFB with different flow fields (at current densities of 50, 100, 150, and 200 mA $\cdot$ cm^{-2} and a flow rate of 3mL $\cdot$ min^{-1} $\cdot$ cm^{-2}).

2. As depicted in Fig. 2(b), the computational domain for the model consists of a flow channel

and a porous electrode.

3. Fig. 13 indicates that the two provinces, Khorasan Razavi and South Khorasan, have the highest proportion of areas categorized as very suitable for establishing small wind plants, while Ilam and Hamedan showed the lowest potential and the lowest share of areas in the very suitable class.

4. The coulombic, voltage, and energy efficiencies of the VRFB with different flow fields are summarized in Fig. 5(b) and (c).

5. Fig. 16 illustrates the locations of adjacent cities to suitable regions for establishing wind power plants for different scenarios of risk.

6. Finally, eight candidates with the highest number of votes were selected as the optimal solutions to the design problem of flow fields, the binary images of which are displayed in Fig. 4(d).

Verbs	
Tense	
Voices	
Sentence 2	

Task 5.5   Decide whether the following sentences are indicative or informative statements. Write "Indi." for an indicative statement or "Info." for an informative statement in the blank before each sentence.

_____ 1. As depicted in Fig. 2(b), the computational domain for the model consists of a flow channel and a porous electrode.

_____ 2. Fig. S13 compares the statistical distribution of the first two principal components of these 1164 flow field images and all the 11 564 samples, indicating that the simulated dataset is sufficiently representative of the search library.

_____ 3. As shown in Fig. S12, the geometry of the flow channel is obtained by extruding the "channel" subgraph of the 2-D binary image by a height of 2 mm (*i.e.*, the channel depth is 2 mm), and the geometry of the porous electrode is represented by a $36 \times 36 \times 1$ mm^3 cuboid (*i.e.*, the electrode thickness is 1 mm).

_____ 4. Fig. 6 illustrates different criteria maps used for selecting suitable sites for establishing wind power plants.

_____ 5. The confusion matrix of the voting result and the corresponding recall, precision, and f_1-score are shown in Tables 3 and 4, respectively.

_____ 6. Figs. 12 and 13 show areas of the very suitable class of regions in the province for establishing large and small solar power plants by different ORness values.

_____ 7. As shown in Fig. 4(c), regeneration processing over 10 cycles had little effect on the filtration properties.

_____ 8. Fig. 6 shows the oscillograms of the output parameters of the micro-HPP generator operating with and without a gearbox.

_____ 9. As shown in Fig. 5(c), no significant fluorescence signal of the nanodots was observed when testing new masks.

_____ 10. Fig. 11 depicts areas of different classes of site suitability by different degrees of decision-making risk for establishing a large wind power plant.

Task 5.6 Read the excerpt from Sample Article 3 (SA3) and then answer the questions below.

Human Activities Changed Organic Carbon Transport in Chinese Rivers during 2004–2018

4. Results

4.1. Spatial OC transport in different rivers

1 Riverine DOC transport indicated by COD (Section 3.1) presented a spatial pattern of "high in the north and low in the south". **2** Excluding the highly polluted Fen River with extremely high NH+ 4-N and COD values of 12.64 mg/L and 58.91 mg/L (Fig. 1, Table

1), respectively, there was a positive linear relationship between the climatological mean COD and the latitude of the sampling sites, with $r = 0.38$ and $p < 0.05$ (Fig. 4a). **3** The COD contents in rivers north of 30°N (5.39 ± 3.66 mg/L) were significantly higher (*t-test, p* < 0.01) than those south of 30°N (2.39 ± 1.14 mg/L). **4** Across different rivers, the climatological mean COD ranged from 1.35 to 16.8 mg/L, with a mean ± std. of 5.16 ± 9.02 mg/L (Table 1). **5** Moreover, high COD contents were generally found for rivers influenced by strong human activities, denoted by high UrbanPer in the catchments (Fig. 4a).

6 The riverine POC concentration was generally "high in the west and low in the east." **7** Excluding the Wei River flowing across the intensely eroded Loess Plateau, China (Fig. 1, Table 1), the climatological mean POC was negatively related to the longitude of the sampling sites, with $r = -0.42$ and $p = 0.06$ (Fig. 4b). **8** The mean POC concentrations were 1.96 ± 1.27 mg/L and 1.34 ± 1.19 mg/L for rivers located west and east of 115°N, respectively. **9** The climatological mean POC ranged from 0.27 mg/L to 4.48 mg/L, with a mean ± std. of 1.69 ± 1.27 mg/L (Table 1). **10** High POC contents were usually found in rivers that originated from highlands (Fig. 4b). **11** For the Liao River with high cropland coverage in the catchment (44.15%), the POC concentration also reached as high as 4.48 mg/L (Figs. 3b, 4b).

4.2. Annual and seasonal variations in OC transport

12 Although there was no obvious spatial pattern, riverine DOC transport indicated by COD showed an annual decrease in most rivers. **13** During 2004–2018, 60.98% (25 of 41) of the studied rivers showed COD decreases, with 41.46% being significant; in contrast, 39.02% of the rivers showed COD increases, with only 21.95% being significant (Fig. 4c). **14** Interestingly, COD decreased mainly in rivers with high COD concentrations and vice versa (Figs. 4a, 4c). **15** The COD concentrations in rivers with decreased and increased COD exhibited a noticeable difference, with mean values of 4.67 ± 3.62 mg/L and 2.53 ± 0.98 mg/L, respectively (*t-test, p* < 0.05). **16** The highly organic-polluted Fen River (Table 1) had the most significant COD decrease during 2004–2018, with an annual decrease of 0.81 mg/L ($p < 0.01$). **17** Being different from COD changes, however, UrbanPer in all catchments increased significantly (Fig. 4c).

…

Questions:

1. What information elements are included in this *Results* section?
2. What is the function of Sentence 1? Could you find other sentences of the same function as Sentence 1?
3. Can you find any sentences that only present specific results but do not indicate the graphics? Why do the writers do so?
4. What verb tense is used to present the major results? Is it true of the writing in your own field?

Briefly commenting on the results (optional)

Sometimes, you may find comments on the results in the *Results* section, especially when partial or middle results were obtained during the research process. The common ways to discuss the results include interpreting the results (IE4), comparing the results with those of other studies (IE5), giving possible reasons (IE6), and assessing the results (IE7).

In a stand-alone *Results* section, such comments are usually very brief, being just a couple of sentences coming right after the description of the result. Extensive and profound discussion will appear in the *Discussion* section. Of course, it will be another story if you are writing a combined *Results and Discussion* section. In such a case, you can elaborate on your comments whenever necessary.

There are voices insisting on no comment in the *Results* section. Such a dispute may derive from disciplinary or individual differences. In your actual writing, you need to make your decision based on careful and purposive considerations. If brief comments help readers achieve a better understanding of the results, it may be good to have them; if the same or similar comments will be repeated in the *Discussion* section, they may not need to appear in the *Results* section. A good principle is to refer to published papers on a topic similar to yours in your target journals.

Task 5.7 The text below is part of the *Results* section of Sample Article 2 (SA2). Analyze what information element(s) each sentence indicates, identify the ways to discuss the results, and then complete the table below.

The Site Selection of Wind Energy Power Plant Using GIS-Multi-Criteria Evaluation from Economic Perspectives

5. Results

…

1 Fig. 11 depicts areas of different classes of site suitability by different degrees of decision-making risk for establishing a large wind power plant. **2** As the results show, reductions in *ORness* correlate with reductions in the suitable classes (suitable and very suitable) area, while areas in the unsuitable classes (unsuitable and very unsuitable) increase. **3** Stated differently, since optimal locations, in the case of lower *ORness* value, corresponding to the highest values in all criteria, the overall area size for the suitable classes (suitable and very suitable) decreases. **4** This may also be understood as a very low-risk investment and is convenient for risk-aversive behaviors (or individuals). **5** Conversely, when *ORness* is high (maximum=1), regions showing high values in even one criterion are selected as optimal locations, indicating a high-risk investment.

…

6 Table 5 lists the optimal purchasing price for small and large power plants with respect to the geographical location of cities. **7** As the figures suggest, the overall price range for electricity varies from 0.074 to 0.384 US$, with Ahar, Taibad, Rashtkhar, Torbat-e Jam, and Asadiyeh contributing to the lowest prices in order of magnitude, while Maragheh, Sarbaz, Arak, Rabor, and Galmurti corresponded to the highest prices, respectively. **8** In other words, cities with lower wind speed conditions charge higher prices for electricity generated by wind power plants since investors tend to choose economically affordable locations.

…

Sentence No.	IE
1	
2	
3	

Sentence No.	IE
4	
5	
6	
7	
8	

LANGUAGE CONVENTION

 Read the following part of the *Results* and *Discussion* section of Sample Article 1 (SA1) and answer the questions below.

Machine Learning-Assisted Design of Flow Fields for Redox Flow Batteries

3. Results and discussion

3.5 Experimental validation

...

1 Fig. 5(a) shows the charge–discharge curves of the VRFB with different flow fields (at current densities of 50, 100, 150, and 200 mA · cm^{-2} and a flow rate of 3mL · min^{-1} · cm^{-2}). **2** Results suggest that the battery with the new flow fields has a comparable electrolyte utilization than that with the IFF while having lower charge and discharge overpotentials and a higher capacity than that with the SFF. **3** At the current density of 100 mA · cm^{-2}, for example, the battery with the HTS04 flow field yields an electrolyte utilization of 78.3%, 10.8% higher than that with the SFF, indicating the important role of designing flow fields in reducing the use of precious electroactive species and the capital cost of VRFBs. **4** The coulombic, voltage, and energy efficiencies of the VRFB with different flow fields are summarized in Fig. 5(b) and (c). **5** Results show that the coulombic efficiency of the battery with all the five flow fields is higher than 95% [Fig. 5(b)], demonstrating the reduced side reaction and low crossover rate of our battery setup. **6** Note that the coulombic efficiency of the battery with the SFF is slightly higher than that of the battery with the other four flow fields due to its decreased charge–discharge duration. **7** Moreover, consistent with the charge–discharge curves, the battery with the new flow fields delivers a higher voltage efficiency than that with the two benchmarks [Fig. 5(b)]. **8** Additionally, the battery with all the flow fields yields a stable energy efficiency as the cycle number increases and the current density varies [Fig. 5(c)]. **9** Since the energy efficiency is the product of the coulombic efficiency and the voltage efficiency, the battery with the new flow fields shows a higher energy efficiency as well. **10** For example, at the current density of 50, 100, 150, and 200 $mA \cdot cm^{-2}$, the battery with the HTS01 flow field exhibits an energy efficiency of 90.1%, 85.7%, 80.7%, and 76.0%, which is 2.7%, 5.3%, 7.3%, and 10.9% higher than that with the SFF and 0.5%, 1.85%, 3.18%, 4.69% higher than that with the IFF, respectively. **11** Fig. 5(d) compares the polarization curves of the VRFB with different flow fields (at a flow rate of 3 $mL \cdot min^{-1} \cdot cm^{-2}$), suggesting that the battery with the new flow fields yields lower polarization losses than that with the SFF and IFF, in agreement with the results of the charge–discharge test. **12** Furthermore, it is shown that using the new flow fields brings about up to 22% increase in limiting current density, from 900 to 1100 $mA \cdot cm^{-2}$, indicating a considerable improvement in the transport of active species within porous electrodes.

13 To further explain the above experimental results, the multiphysics simulation described in Section 3.2 was performed. **14** Essentially, the superior performance of the battery with the new flow fields can be attributed to improvement in mass transfer and the resulting reduction in concentration polarization losses. **15** As shown in Fig. 5(e), the new flow fields yield higher z-component electrolyte velocity u_z within the electrode, indicating the stronger under-rib convection. **16** For example, the maximal Uz across the HTS01 flow field reaches 0.17 $cm \cdot s^{-1}$, higher than the SFF (0.04 $cm \cdot s^{-1}$) and the IFF (0.16 $cm \cdot s^{-1}$). **17** As a result, using the new flow fields can distribute electroactive species more uniformly throughout the electrode (Fig. 5(f)), which is corroborated by the higher uniformity factor of the new flow fields (0.821–0.827) than the SFF (0.731) and the IFF (0.818). **18** Since the improved species transport increases the local concentration of reactants within the electrode, the concentration loss as well as the total overpotential can be significantly decreased [Fig. 5(g)]. **19** For example, the battery with the HTS01 flow field suffers from an overpotential of 25.5 mV, which is 20% lower than that with the SFF (31.9 mV) and slightly lower than the IFF (26.8 mV).

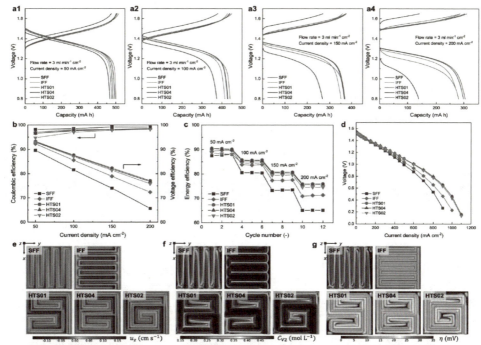

Fig. 5 Experimental validation. (a) Charge–discharge curves of the battery with different flow fields (SFF, IFF, HTS01, HTS04, and HTS02) at an area-specific flow rate of 3 ml min^{-1} cm^{-2} and current densities of 50 mA cm^{-2} (a1), 100 mA cm^{-2} (a2), 150 mA cm^{-2} (a3), and 200 mA cm^{-2} (a4). (b) Coulombic efficiency and voltage efficiency of the battery with different flow fields as a function of current density. (c) Energy efficiency of the battery with different flow fields as a function of charge–discharge cycle numbers. (d) Polarization curve of the battery with different flow fields at an area-specific flow rate of 3 ml min^{-1} cm^{-2}. (e)–(g) Simulated distribution of the z-component electrolyte velocity (u_z) (e), V^{2+} concentration ($c_{V^{2+}}$) (f), and overpotential (η) (g) across an xy-plane (near the channel/electrode interface) inside the porous electrode.

Questions:

1. What information elements have you found in this sample?
2. Is the figure title a complete sentence or a phrase?
3. What is the figure legend mainly about?
4. Is the figure legend expressed in phrases or complete sentences?
5. What verb tenses are used in the text? Why are they used?

Using visuals

The *Results* section of a research article is characterized by the use of graphics, including tables, graphs, photos, etc., with the corresponding text mainly reporting the data in the graphics or the findings observed from the data obtained. The purpose of using many graphics in the *Results* section is to provide information about the major research findings visualized. In a paper, all the tables are put into the category of "Table" and numbered according to the sequence they appear. Other graphics, including statistical graphs, pictures, images, and sketches, are categorized as "Figure." Statistical tables and graphs are used in different cases. Table 5.1 shows the most useful cases for using tables and graphs.

Table 5.1　The choice between data displayed in figures or tables.

Most useful	Table	Figure
When working with	number	shape
When concentrating on	individual data values	overall pattern
When accurate or precise actual values are	more important	less important

(From *Writing Scientific Research Articles* by Margaret Cargill and Patrick O'Connor)

Each table or figure should have a title or caption, containing a brief description of the key information being presented. The table title is above the table while the figure title is below the figure. The caption of either a table or a figure goes under it. In one paper, tables and figures are numbered separately. The number of tables and figures is usually highlighted in bold.

Tables are often formatted as having only three horizontal lines: a top line, a line between the column headings and the listed items, and a bottom line (e.g. Table 5.1). Sometimes, supplementary lines are added to better present information. Vertical lines are usually not used.

Figures may vary substantially in terms of subject field. They could be images from procedures such as microscopy, pictures demonstrating chemical formulas, or sketches showing equipment or process. In addition, statistical graphs are often employed to present research results, and the common ones are column or bar charts, line charts, pie charts, radar charts, histograms, scatter diagrams, etc. Figures may have figure legends, which usually contain five possible parts:

- A title summarizing what the figure is about;
- Details of results or models shown in the figure or supplementary to the figure;
- Additional explanation of the components of the figure, the method used, or essential details of the figure's contribution to the entire *Results* section;
- Description of the units or statistical notation included;
- Explanation of any other symbols or notation used.

Task 5.8 Read the figure legends of the following figures or tables from sample articles and analyze the contained parts.

Sample article	Figures or tables	Parts contained in the figure legends
1	Table 1	
3	Fig. 4	
3	Fig. 7	
4	Table 2	
9	Fig. 3	

Task 5.9 Read the excerpt from the *Results* section of Sample Article 2 (SA2) and answer the questions below.

> **The Site Selection of Wind Energy Power Plant Using GIS-Multi-Criteria Evaluation from Economic Perspectives**
>
> **5. Results**
>
> …
>
> 1 Table 6 shows the optimal purchasing price for electricity obtained from large power

plants, with respect to different alternatives. **2** Evidently, 84 options were categorized as very suitable for large wind turbine establishments, with the lowest price estimated at 0.047US$, and the highest price at 0.182 US$. **3** These prices appear to drop as wind speed increases in a region and vice versa.

Table 6
Optimal purchase price of electricity generated by large power plants with respect to the geographical location of alternatives.

Id	Wind speed (m.s^{-1})	Optimal price (US$)	Id	Wind speed (m.s^{-1})	Optimal price (US$)
1	5.87	0.125	43	7.19	0.069
2	6.33	0.101	44	6.33	0.101
3	6.16	0.109	45	6.72	0.085
4	5.47	0.156	46	6.35	0.099
5	6.72	0.084	47	6.73	0.084
6	5.47	0.156	48	6.25	0.104
7	5.74	0.135	49	5.99	0.119
8	5.96	0.121	50	7.78	0.0545
9	6.12	0.112	51	7.53	0.061
10	6.39	0.098	52	6.84	0.081
11	6.78	0.082	53	5.74	0.135
12	6.29	0.102	54	6.47	0.094
13	7.34	0.064	55	7.27	0.066
14	7.03	0.073	56	6.69	0.088
15	6.42	0.096	57	5.97	0.121
16	5.97	0.121	58	5.39	0.163
17	6.02	0.117	59	8.12	0.047
18	6.88	0.078	60	6.23	0.106
19	6.25	0.104	61	5.92	0.123
20	6.68	0.085	62	7.88	0.052
21	6.34	0.102	63	6.38	0.098
22	7.24	0.067	64	6.27	0.1041
23	5.74	0.135	65	5.81	0.131
24	7.86	0.052	66	6.31	0.101
25	6.62	0.088	67	6.08	0.114
26	5.91	0.124	68	6.95	0.0765
27	7.23	0.067	69	5.89	0.125
28	7.16	0.069	70	6.82	0.083
29	5.31	0.172	71	5.82	0.131
30	5.99	0.119	72	6.08	0.113
31	7.27	0.066	73	5.77	0.133
32	5.42	0.161	74	6.28	0.103
33	5.35	0.167	75	5.21	0.182
34	7.11	0.071	76	6.14	0.112
35	7.72	0.055	77	6.55	0.091
36	5.89	0.125	78	5.82	0.131
37	7.22	0.068	79	6.47	0.094
38	5.72	0.137	80	5.93	0.123
39	5.61	0.145	81	8.53	0.041
40	6.12	0.111	82	5.71	0.138
41	6.58	0.09	83	6.81	0.081
42	8.87	0.036	84	5.31	0.171

Questions:

1. Which parts of the text can be directly observed in the table? Underline them.
2. What other information is not directly observed in the table but provided in the text? What is its function?
3. How is the table formatted?
4. Can Sentence 3 be omitted? Why or why not?
5. Are there any other ways to indicate the table?
6. Can Sentence 1 be omitted? Why or why not?

Describing changes, comparison, and relationships

Three ways are commonly used to report research findings: comparing variables, data, values, conditions, etc., showing the fluctuation of variables over time, and revealing the relationships among variables. Each way is characterized by its distinctive language expressions.

To compare variables, data, values, or conditions, you can use comparative or superlative expressions such as *higher, lower, more... than, the highest, the most, to compare, compared with...*, etc. There are other forms showing comparison, for example, *as...as, not as (so)... as, similar to,* and *like*. When making comparison, make sure what are compared should belong to the same category; in other words, what you compare should be comparable.

Frequently used verbs to show the fluctuation of variables include *increase, decrease, rise, decline,* etc. Expressions such as *be reduced, be up- or down-regulated,* and *be promoted* are used to indicate both the changes of variables and the causes of the changes.

As for the relationships among variables, typical phrases such as *be related to, be associated with, correlate to/with,* etc. are used. Adverbs such as *positively, negatively, significantly, closely, highly, slightly,* etc. are often adopted to indicate the direction or strength of the relationship.

Task 5.10   Read the excerpt from the *Results* section of Sample Article 3 (SA3). Underline the expressions suggesting the three different ways of reporting research findings and then complete the table below.

> **Human Activities Changed Organic Carbon Transport in Chinese Rivers during 2004–2018**
>
> **4. Results**
>
> *4.2. Annual and seasonal variations in OC transport*
>
> **1** Although there was no obvious spatial pattern, riverine DOC transport indicated by COD showed an annual decrease in most rivers. **2** During 2004–2018, 60.98% (25 of 41) of the studied rivers showed COD decreases, with 41.46% being significant; in contrast,

39.02% of the rivers showed COD increases, with only 21.95% being significant (Fig. 4c). **3** Interestingly, COD decreased mainly in rivers with high COD concentrations and vice versa (Figs. 4a, 4c). **4** The COD concentrations in rivers with decreased and increased COD exhibited a noticeable difference, with mean values of 4.67 ± 3.62 mg/L and 2.53 ± 0.98 mg/L, respectively (*t-test, p* < 0.05). **5** The highly organic-polluted Fen River (Table 1) had the most significant COD decrease during 2004–2018, with an annual decrease of 0.81 mg/L ($p < 0.01$). **6** Being different from COD changes, however, UrbanPer in all catchments increased significantly (Fig. 4c).

7 The mean annual POC concentrations in most rivers decreased during 2004–2018. **8** Of the studied rivers, 90.48% (19 of 21) showed POC decreases, with 42.86% being significant (Fig. 4d). **9** Specifically, the Han River, a tributary of the Yangtze River (Fig. 1, Table 1), had the most significant POC decrease; the mean POC concentration was 3.61 mg/L in 2005 but only 1.36 mg/L in 2018, with an approximately 60% decrease during 2004–2018. **10** Along with riverine POC decreases, vegetation coverage indicated by NDVI increased in most regions, especially in northeastern China and on the Loess Plateau (Figs. 1, 4d). **11** Moreover, for the two rivers with POC increases (Neng and upper Yangtze rivers, Table 1), NDVI increases in corresponding catchments were negligible, with *r* values of 0.07 and 0.06 ($p > 0.05$), respectively (Fig. 4d).

12 Both COD and POC transport also showed apparent seasonal variations. **13** Across different rivers, seasonal variations in COD transport were not similar; the months with the highest COD contents were different (Fig. S2). **14** Overall, both monthly mean COD and NH_4^+-N in all studied rivers had higher values in winter than in summer (Fig. 2c). **15** However, the monthly mean POC in all rivers showed a unimodal distribution, with the highest POC contents in summer (Fig. S2). **16** Seasonal variations in POC transport were significantly determined by basin precipitation; there was a significant linear relationship between monthly climatological POC and rainfall, with $r = 0.96$ and $p < 0.05$ (Fig. 2d).

Ways to report research findings	Expressions
Comparing variables, data, conditions, etc.	
Showing the fluctuation	
Revealing the relationships among variables	

Using appropriate tense

Different verb tenses are used to write the *Results* section. The conventional verb tense used for writing each element of this section is shown in Table 5.2.

Table 5.2 Verb tenses for writing different elements of the *Results* section

Information element	Verb tense
Re-mentioning methods-related details	past tense
Indicating graphics	present tense
Highlighting important results	past tense
Brief comments	present tense or modal auxiliaries

It is worth mentioning that in some research fields such as economics and engineering. The present tense can be used to report the major findings. Exceptions like this do exist owing to wide differences across disciplines and journals. Sometimes, the past tense may also be used to make comments. Detailed information on this will be given in the *Discussion* section.

Task 5.11 Read the following extract from the *Results* section of Sample Article 4 (SA4) and fill in the blanks with the appropriate verb tenses and voices.

A Deep Learning System to Screen Novel Coronavirus Disease 2019 Pneumonia

4. Results

…

4.3.3. Classification for a single image patch

1 A total of 1710 image patches (1) _____ (acquire) from 90 CT samples, including 357 COVID-19, 390 IAVP, and 963 ITI (ground truth). **2** To determine the optimal approach, the design of each methodology (2) _____ (assess) using a confusion matrix. **3** Two classification models (3) _____ (evaluate): with and without the location-attention mechanism, as shown in Tables 1 and 2.

4 The average f_1-score and the overall accuracy rate for the two models (4) _____ (be) 0.750/0.764 and 78.5%/79.4%, respectively. **5** Furthermore, the location-attention mechanism (5) _____ (use) to improve the respective accuracy rate of the COVID-19 and IAVP groups, and (6) _____ (show) to result in a remarkable improvement of 5.0% (260/273) and 1.4% (276/280). **6** This evidence (7) _____ (indicate) that the second model with the location-attention mechanism achieved better performance. **7** Therefore, that model (8) _____ (use) for the rest of this study.

8 Moreover, as the ITI group (9) _____ (use) to remove disturbing factors in this study, it (10) _____ (ignore) and not counted by the Noisy-OR Bayesian function in the final step. **9** To retain consistency in the next steps, we further (11) _____ (compare) the average f_1-score and accuracy rate for the first two groups, which (12) _____ (be) 0.720 and 74.0%, respectively.

Characteristic expressions

The characteristic expressions used in the *Results* section can be classified into the groups below:
- Verbs used in indicating graphics;
- Expressions comparing variables, data, values, conditions, etc.;
- Expressions indicating the fluctuation of variables;
- Expressions showing the relationships among variables;
- Statistical expressions.

Scan the QR code for a list of the expressions. Try to get yourself familiar with these expressions and be ready to use them in your future writing.

Task 5.12 Complete the following sentences with the expressions provided in the box below.

| is given | are directly correlated | are summarized | higher | illustrates |
| demonstrated | more suitable | std. | highest | reduced |

1. The radiological findings _____ in Table 3.

2. This is the reason why ~20 at.%Fe alloys exhibit slowest diffusion, as _____ in Fig. 1a.

3. Fig. 13 indicates that the two provinces, Khorasan Razavi and South Khorasan, have the _____ proportion of areas categorized as very suitable for establishing small wind plants, while Ilam and Hamedan showed the lowest potential and the lowest share of areas in the very suitable class.

4. ORness values _____ with the degree of investment risk; i.e., increased ORness corresponds to higher risk, and vice versa.

5. The climatological mean POC ranged from 0.27 mg/L to 4.48 mg/L, with a mean ± _____ of 1.69 ± 1.27 mg/L (Table 1).

6. The relative proportion of areas in the very suitable class to the total area of the province _____ in percent in Fig. 15 for small wind power plants.

7. Fig. 16 _____ the locations of adjacent cities to suitable regions for establishing wind power plants for different scenarios of risk.

8. In our experiments, we found that most of the regenerated surgical masks had a PFE much _____ than 30%, with an average value of 92.3%.

9. After the first-round screening, the original search space (~10^4) was dramatically _____ to a human-tractable range (~10^2).

10. Compared with other proposed decontamination methods for masks, such as organic solvent, ultraviolet (UV) radiation, and hydrogen peroxide steam, hot water decontamination is _____ for people to perform at home without the use of additional solvents or high-tech equipment.

Check your understanding

Task 5.13 Choose two from the sample articles (SA1–SA12) in this textbook and read the *Results* sections. Do the following activities.

1. Find what information elements are included in the *Results* section by answering the questions below:
    - Is the *Results* section an independent section or combined with the *Discussion* section?
    - Could you find any method-related details in the section? If yes, mark them.
    - Underline all the sentences where graphics are indicated. How is the table or figure indicated? Is it expressed in a sentence or simply placed in parentheses after highlighting important results or both?
    - Underline the sentences where major findings are reported. Does the report focus on comparison among groups, fluctuation of variables over time, or relationships among variables?
    - Could you find any comments on the findings? If yes, are they brief or extensive? What kind of comment do they belong to, interpreting the results, comparing results with other studies', giving reasons for the results, or assessing the results?

2. Get familiar with the language features in the *Results* section by answering the questions below.
    - Observe the subheadings in the *Results* sections. Are they full sentences or phrases?
    - What verb tenses are used for writing the information elements and figure legends? Find two examples for each case.
    - Are there comparative (-er, more) or superlative (-est, most) expressions or other expressions of comparison (as...as, not so/as…as)? If yes, highlight them.
    - Could you find expressions indicating the fluctuation of variables (e.g. *increase, decrease* etc.) or showing the relationships among variables (e.g. *associated, correlated,* etc.)? If yes, highlight them.
    - Are there statistical expressions such as *statistically significant, no significant differences were found*, etc.? If yes, highlight them.

Task 5.14 Pick two journal articles in your field. Analyze the *Results* section and provide samples for the items listed in the table below. If there is no corresponding information for a certain item, write "None" in the blank.

Item	Article 1	Article 2
Article title		
Heading of the *Results* section		
Subheadings		
Re-mentioning method-related details		
Indicating graphics		
Comparison among the groups		
Fluctuation of the variables		
Relationships among the variables		
Statistical treatment of the data		
Interpreting the results		
Comparing the results with those from other studies		
Giving reasons for the results		
Assessing the results		

Unit task

Interpreting the Table

The table below is from the *Results* section of one of our sample articles in this textbook. Try to interpret the data and write a corresponding text of about 200 words. Refer to what you have learned in this unit. Include the information elements and follow the language conventions in your writing.

Table 5 Optimal purchase price of electricity generated by small power plants based on the geographical location of cities.

Id	Name	Wind speed (m.s^{-1})	Optimal price (US$)	Id	Name	Wind speed (m.s^{-1})	Optimal price (US$)
1	Parsabad	5.49	0.154	48	Taybad	6.84	0.081
2	Poldasht	4.35	0.311	49	Bajestan	5.27	0.175
3	Meshginshahr	6.08	0.113	50	Jandaq	5.11	0.192
4	Salmas	4.69	0.248	51	Eshqabad	4.43	0.294
5	Urmia	4.71	0.245	52	Qaen	5.76	0.134
6	Astara	4.81	0.231	53	Seh Qaleh	5.34	0.168
7	Lotfabad	4.54	0.274	54	Ardestan	4.22	0.341
8	Marand	5.14	0.189	55	Anarak	4.51	0.281
9	Ahar	7.01	0.074	56	Alavijeh	4.32	0.317
10	Tabriz	5.16	0.186	57	Naein	4.57	0.268
11	Varzaqan	4.84	0.226	58	Isfahan	4.3	0.321
12	Naqadeh	4.53	0.276	59	Asadiyeh	6.15	0.111
13	Bojnord	5.36	0.167	60	Birjand	6.09	0.113
14	Shirvan	4.72	0.243	61	Shush	4.95	0.211
15	Quchan	4.62	0.261	62	Rofayyeh	4.21	0.342
16	Jajarm	4.43	0.293	63	Nehbandan	5.68	0.141
17	Sarakhs	4.64	0.257	64	Yazd	4.25	0.333
18	Sari	4.12	0.365	65	Bafq	4.38	0.305
19	Maragheh	4.05	0.384	66	Izad Khast	4.86	0.223
20	Mianeh	5.02	0.203	67	Ravar	4.62	0.259
21	Zanjan	5.89	0.125	68	Mal-e Khalifeh	4.41	0.298
22	Azadshahr	4.21	0.343	69	Yasuj	4.71	0.246
23	Tarom	4.95	0.211	70	Zabol	6.41	0.097
24	Qazvin	5.61	0.145	71	Abarkuh	4.42	0.295
25	Sardasht	4.36	0.309	72	Razi	4.12	0.366
26	Saqqez	4.65	0.255	73	Rafsanjan	5.04	0.211
27	Marivan	4.48	0.285	74	Kerman	4.76	0.237
28	Sanandaj	4.54	0.273	75	Shahr-e Babak	4.17	0.353
29	Mashhad	5.04	0.199	76	Abadan	4.91	0.216
30	Neyshabur	5.02	0.203	77	Nosratabad	4.85	0.225
31	Sabzevar	4.89	0.219	78	Manujan	4.45	0.291
32	Shahroud	5.25	0.177	79	Sirjan	4.18	0.351
33	Damghan	5.26	0.176	80	Rabor	4.09	0.374
34	Semnan	4.91	0.217	81	Bam	4.36	0.308
35	Tehran	4.35	0.311	82	Mohammadabad	4.35	0.311
36	Zarrin Rood	5.82	0.131	83	Zahedan	5.03	0.201
37	Nahavand	4.87	0.222	84	Kaki	4.92	0.215
38	Tuyserkan	4.85	0.224	85	Khash	4.33	0.314
39	Qom	4.41	0.298	86	Galmurti	4.14	0.359
40	Hamadan	4.27	0.328	87	Bandar Lengeh	4.35	0.311
41	Sonqor	4.71	0.247	88	Saravan	4.29	0.324
42	Harat	4.61	0.261	89	Iranshahr	4.39	0.303
43	Arak	4.07	0.378	90	Sarbaz	4.06	0.382
44	Torbat-e Jam	6.14	0.111	91	Qasr-e-qand	4.76	0.237
45	Gonabad	5.71	0.138	92	Pishin	4.62	0.259
46	Kashmar	5.47	0.156	93	Chabahar	5.29	0.172
47	Roshtkhar	6.29	0.103				

Unit 6

Discussing Your Study

Learning objectives

In this unit, you will:
- understand the general function of the *Discussion* (& *Conclusion*) section;
- learn about the common information elements in the *Discussion* (& *Conclusion*) section;
- develop linguistic strategies for writing an effective *Discussion* (& *Conclusion*) section.

Self-evaluation

Read through the *Discussion* section of Sample Article 2 (SA2) "The Site Selection of Wind Energy Power Plant Using GIS-Multi-Criteria Evaluation from Economic Perspectives" and answer the following questions.
- How do the authors start this section?
- What sentences are used to elaborate on the authors' observed results?
- What sentences are used to give reasons for their research setting, results, or procedures?
- What sentences are used to relate their study to others' studies?
- How do the authors end this section?

A research paper normally ends by discussing the study and its findings. The various arrangements of the *Results*, *Discussion*, and *Conclusion* sections have been explained in Unit 2. For pedagogical simplicity, this unit addresses the discussion as an independent section and the conclusion as the ending part of the discussion.

In the *Discussion* (& *Conclusion*) section, authors step back and take a broad look at their findings and the entire work. As the final section, the *Discussion* section should bring everything together by highlighting the major results, relating the present study to existing works in the field, addressing the proposed research questions or hypotheses, and stating the significance or contribution of the work. Some fundamental questions should be addressed in this section.

- What is/are the most significant finding(s)?
- What does a specific result indicate?
- How do the findings provide solutions to the posed problem(s)?
- How is the study compared with other related research?
- What universally applicable conclusions could be drawn?
- What are the strengths and limitations of the study?
- How does the study contribute to the area?

For novice researchers, the *Discussion* (& *Conclusion*) section may be the most difficult to write. This unit will help you understand and grasp some strategies for writing an effective *Discussion* (& *Conclusion*) section.

INFORAMTION CONVENTION

 Below is the *Discussion* section of Sample Article 4 (SA4). Please read it and specify what each paragraph is about.

A Deep Learning System to Screen Novel Coronavirus Disease 2019 Pneumonia

Discussion

1 **1** COVID-19 has caused serious public health and safety problems, and hence has become a global concern [33–35]. **2** In the early stage of COVID-19, some patients may already have positive pulmonary imaging findings but no sputum or negative nucleic acid testing results from sputum or nasopharyngeal swabs. **3** These patients are not diagnosed as suspected or confirmed cases. **4** Thus, they are not isolated or treated in a timely manner, making them potential sources of infection. **5** Meanwhile, CT examination is routinely performed on every patient with fever and respiratory symptoms in the early stage, and is repeated for dynamic observation, since it is cheap and easy to operate. **6** Using CT images to screen patients can improve the early detection of COVID-19, and ease the pressure on laboratory nucleic acid testing.

sputum 痰

nasopharyngeal swab 鼻咽拭子

respiratory symptoms 呼吸系统症状

2 **7** The CT imaging of COVID-19 presents several distinct manifestations, according to previous studies [29,30,36]. **8** These manifestations include focal ground-glass shadows mainly distributed along the pleura, multiple consolidation shadows accompanied by the "halo sign" of the surrounding ground-glass shadow, multiple consolidations of different sizes, and grid-shaped high-density shadows. **9** An experienced radiologist can make judgments on the possibility of COVID-19 based on his or her clinical experience; however, such judgments are easily influenced by subjective factors and individual proficiency. **10** In comparison, deep-learning system-based screen models reveal more specific and reliable results by digitizing and standardizing the image information. **11** Hence, they can assist physicians in making a clinical decision more accurately.

pleura 胸膜

grid-shaped 网格状的

3   12 There have been many precedents for artificial intelligence (AI)-assisted models that are now widely used in clinical practice, such as the pulmonary nodules diagnostic system. 13 In June 2019, Ardila et al. [20] proposed a deep learning algorithm that used a patient's current and prior CT volumes to predict the risk of lung cancer. 14 This model achieved an accuracy rate of nearly 94.4% on 6716 cases, and performed similarly on an independent clinical validation set of 1139 cases. 15 In addition to having a high accuracy rate, an AI-assisted model can do the work faster and more efficiently than a human.

4   16 In this study, deep learning technology was used to design a classification network for distinguishing COVID-19 from IAVP. In terms of the network structure, the classical ResNet was used for feature extraction. 17 A comparison was made between models with and without an added location-attention mechanism. 18 The experiment showed that the aforementioned mechanism could better distinguish COVID-19 cases from others. 19 Furthermore, multiple enhancement methods were involved in our study, such as image patch vote and Noisy-OR Bayesian function, in order to determinate the dominating infection types. 20 All these efforts produced a consistent improvement in the average f_1-score and accuracy rate.

IAVP 甲型流感病毒性肺炎

5   21 This study has some limitations. First, the manifestations of COVID-19 may have some overlap with the manifestations of other pneumonias such as IAVP, organizing pneumonia, and eosinophilic pneumonia. 22 We only compared the CT manifestation of COVID-19 with that of IAVP. 23 A clinical diagnosis of COVID-19 still needs to combine the patient's contact history, travel history, first symptoms, and laboratory examination. 24 Second, the number of model samples was limited in this study. 25 The number of training and test samples should be expanded to improve the accuracy in the future. 26 More multi-center clinical studies should be conducted to cope with the complex clinical situation. 27 Moreover, efforts should be made to improve the segmentation and classification model. 28 A better exclusive model could be designed for training, the segmentation and classification accuracy of the model could be improved, and the generalization performance of this algorithm could be verified with a larger dataset.

Paragraph	What it is about
1	
2	
3	
4	
5	

Overall structure of the Discussion (& Conclusion) section

The contents in the *Discussion* (& *Conclusion*) section may vary from paper to paper, but some basic information elements (IEs) do exist and appear frequently in different disciplinary papers. Fig. 6.1 presents the possible IEs that might be included in this section. These elements can be classified into four information chunks: moving into discussing the study, commenting on key results, evaluating the study, and concluding the study. All these IEs need to be connected with the issues raised in the *Introduction* section.

Not all the information elements listed here will appear in a paper. The arrangement of the information elements shown here is not strictly followed by all authors. Some may appear in cycles. Some may be integrated and not clear-cut from each other. However, the progression from specific to more general information elements is conventional, which is contrary to the information arrangement of the *Introduction* section.

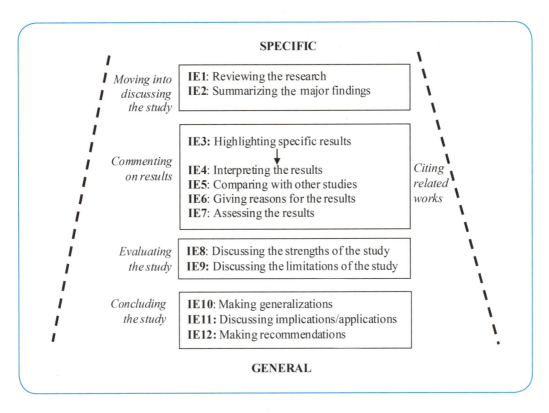

Figure 6.1   Information elements of the Discussion (& Conclusion) section

Task 6.1   Read the *Discussion* section in SA4 and answer the following questions.

1. Does the structure of the *Discussion* section present an upside-down funnel shape? How is it organized from specific to general?
2. Why do the authors review the results of previous studies in the *Discussion* section?
3. Does the first paragraph follow the specific-to-general pattern? If no, what pattern does it follow?
4. Could you use one sentence to summarize the function of the first three paragraphs?
5. Why do the authors report the limitation(s) of the study?
6. Do you think it is necessary to highlight the implication of the study in the *Discussion* section? Why or why not? Please underline the expressions highlighting the implication of the study.

Moving into discussing the study

The *Discussion* (& *Conclusion*) section commonly begins by revisiting some aspects of the research work to transfer from reporting the work to discussing it. This information chunk often contains two information elements.

IE1: Reviewing the research

Although there is no standard way to start the *Discussion* section, many authors tend to begin this section by reminding readers of some background information about the study, such as the research aim, question or hypothesis, the problem the study was designed to solve, important factual information, original prediction, theory, assumption, the rationale for study procedures, and/or major study activities. Below are some examples.

Using three atomistic modelling techniques: µs-scale MD, k-ART and conventional lattice kMC, the vacancy diffusion in Ni-Fe alloys has been studied.

COVID-19 has caused serious public health and safety problems, and hence has become a global concern [33–35].

IE2: Summarize major findings

The *Discussion* section can be started by reviewing the principal findings of the *Results* section. This helps readers further account for what the study has detected and fulfilled. Below are two examples.

We have shown that temporal fluorescence data can be used to precisely localize a set of point fluorescent emitters in a homogeneous heavily scattering environment.

Findings confirmed the possibility of mismatched operation of the generator and the turbine.

In the case that the conclusion of the study stands alone as an independent section, IE1 and/or IE2 may appear there, initiating the *Conclusion* section. In other words, this information chunk may appear at the early stage of either the *Discussion* section or the *Conclusion* section.

Task 6.2 Read the first paragraph abridged from the *Discussion* section of Sample Article 2 (SA2), and answer the questions below.

The Site Selection of Wind Energy Power Plant Using GIS-Multi-Criteria Evaluation from Economic Perspectives

Discussion

1 This study can be divided into two parts (i) determining the optimal locations for the construction of wind farms and (ii) determining the optimal purchase price of electricity generated by wind farms in these areas. **2** In several studies, different dimensions of determining the optimal locations for the construction of wind farms were investigated [76–80]. **3** The focus of these studies is on the use of MCE to determine optimal locations. **4** The accuracy of the results of multi-criteria spatial decision-making systems depends on (i) the comprehensiveness of the effective criteria, (ii) the weight accuracy determined for the effective criteria, and (iii) the model used to combine the values and weight of the effective criteria.

Questions:

1. How do the authors start this section?
2. What is the relationship among Sentences 2–4?
3. In which sentences do the authors use listing? Why do they do so? What grammatical structures are deployed?
4. Could you predict what will be discussed in the next paragraph without reading the original paper?

Commenting on results

> This information chunk, representing the centerpiece of the *Discussion* section, serves the purpose of establishing the meaning and significance of the research results in relation to the relevant field. The conventional ways to comment on results are given below. These elements can serve as a checklist for you as you write. You may not have to take every step for every result you intend to report and discuss, but it is worthwhile thinking about each step as you draft the *Discussion* section. To comment on results, authors prefer using the expressions with modal meanings (such as *may, can, likely*) constructed in the present simple tense, either active or passive form.

IE3: Highlighting specific results

Before making any comments, authors need to let readers know exactly which results they choose to discuss. IE3 serves this purpose. It highlights specific results to be discussed. This element and the other elements in the chunk of commenting on results form result-comment sequences, which are recursive in the *Discussion* section. The element may stand alone as independent sentences or be embedded in the subsequent commenting elements in highly complex texts. The result-comment sequence is repeated, and subheadings are sometimes used to help readers locate the discussion about specific results. Below are some examples.

We also found an exponential relationship between POC(%SPM) and SPM for Chinese rivers entering estuaries; however, POC(%SPM) was lower than the global curve only when with SPM < 100 mg/L (Fig. 8b).

Basin NDVI was significantly negatively correlated with the POC concentration, with r = -0.55 and p < 0.05 (Fig. 6).

IE4: Interpreting the results

Sometimes the results, especially numerical ones, may not be explicit and easy to understand. Authors need to explain what the results mean or indicate to help readers better understand them. See some examples below.

At the fuel temperatures of 60 °C and 90 °C, the fouled injector has a higher degree of collapse, indicating that the existence of the deposits can enhance the flash-boiling process.

Results suggest that the eight promising candidates have lower P1 (18 1), more moderate P2 (2.45 0.16), lower P3 (0.23 0.18), higher P4 (1490 100), and higher P5 (20.1 1.8), compared with the original dataset.

IE5: Comparing with other studies

This element represents a widely adopted way to discuss the study results. The results at hand are compared with those in related works, through which authors confirm or refute existing understandings or findings, demonstrate the validity and reliability of the present study, and visualize how their study is identical to or different from others.

By doing this, authors contextualize the specific findings and assign greater importance to the findings as well as the study. Below are some examples.

The influence of alcohol fuel on the ID is also proved by relevant research [38].

Overall, this research found that except for the flame propagates and the temperature threshold which is mentioned by Liu et al. [28], the equivalence ratio is also important for the second stage of ignition.

The CO emission of D64P16DB20 decreased further when DBE was added to the DP mixture. The trends were similar to those reported by Liu et al. [42].

Note that simple reporting of consistency or inconsistency of the study results with previous findings or existing understandings may not be enough. Preferably, authors should explain or comment on the consistency/inconsistency to demonstrate their prudent attitude and rigorous approach.

IE6: Giving reasons for the results
This element aims to disclose what underlying reasons may account for the results. The study results could be unexpected, unsatisfying, or different from the previous literature. Possible reasons for the problems should be given. Below are some examples.

Due to the worse combustion stability of the optical engine compared to the typical commercial engine, the HCs emission is much higher than the average level, resulting in a higher concentration of nucleation mode particles.

It is worth noting that when 4H2 is above 10%, the P_{max} shows a relatively slow increase. This is because that a small 4H2 can significantly improve the combustion process, while the fuel is relatively less sensitive to the further increasing 4H2.

IE7: Assessing the results
Sometimes, authors evaluate the results based on existing theories, beliefs, or practices.

Data reliability can also be assessed to ensure the validity of the results. Below are some examples.

The present results are significant in at least two major respects.

However, these results were not very encouraging.

Task 6.3 Read the second paragraph excerpted from the *Discussion* section of Sample Article 2 (SA2), and answer the questions below.

The Site Selection of Wind Energy Power Plant Using GIS-Multi-Criteria Evaluation from Economic Perspectives

Discussion

1 Based on our knowledge of previous studies and the opinion of experts, a comprehensive set of criteria, including environmental and economic, was considered in assessing the suitability of areas for the construction of wind farms. **2** The BWM was used to determine the weight of the criteria. **3** This weighting method was used for the first time in determining suitable areas for constructing renewable power plants. **4** In previous studies, AHP, ANP, and Fuzzy methods were used. **5** The BWM has notable advantages over other methods. **6** First, it uses fewer pairwise comparisons, and second, it has a higher consistency ratio [50,52,81]. **7** In previous years, this model was used to determine the weight of effective criteria in the field of waste disposal [82–84], bicycle stations [85,86], hospitals [87], and emergency facility planning [88]. **8** Also, in these studies, various models based on MCE, including WLC, AHP, and fuzzy, have been used to determine the optimal locations. **9** Each of these models has advantages and disadvantages. **10** One of the most important limitations of these models is not considering the concept of risk in decision-making. **11** Therefore, in this study, for the first time, OWA has been used to determine suitable locations for the construction of wind farms. **12** OWA output is the mapping of optimal wind farm locations in various risk-based decision scenarios. **13** In recent years, the efficiency of this model has been confirmed in a number of studies in the fields of landfill [89–91], parking [92–94], and ecotourism [95–97].

Questions:

1. Is it a good idea to delete the first sentence? Why?
2. What is the function of the phrase "based on…" in the first sentence?
3. In which sentences do the authors compare their method with others?
4. Is it necessary to rewrite Sentences 4-5 into one sentence? If so, why? If not, why not?
5. What type of citation is used in this paragraph? What function does each citation perform?
6. What could a mind mapping for this paragraph be like?

Task 6.4 The following sentences are used to comment on results. Read them carefully and determine the information element each sentence represents. Underline the signal words or expressions that help you make your decisions.

1. At the current density of 100 mA·cm^2, for example, the battery with the HTS04 flow field yields an electrolyte utilization of 78.3%, 10.8% higher than that with the SFF, indicating the important role of designing flow fields in reducing the use of precious electroactive species and the capital cost of VRFBs.

2. A surprising fact has been revealed, namely that an increase in the concentration of faster species (Fe atoms) in fcc Ni-Fe decreases the overall atomic diffusion.

3. The main reason is that between 40 MPa and 80 MPa the fuel injection rate is faster with the increase of the injection pressure, and there is more fuel to start burning at the same time which causes the higher maximum HRR.

4. This may indicate that hybrid GPF may have a higher filtration efficiency in rural and motorway: rural driving and motorway driving have higher exhaust temperature than urban because of higher load and less re-start.

5. In this work, vehicle 4 gets similar PN concentration spikes with Yang's study, but the PN concentration spike for vehicle 3 is only about 1.0×10^7 #/cm^3 (except for cold-start period). This means optimized hardware or control strategy could reduce PFI hybrid particle emission significantly even without a particle filter.

6. Table 5 illustrates the emission factor ratio (efk). For Vehicle 1 and Vehicle 2, efu = 3.25 and 2.94; for Vehicle 3 and Vehicle 4, efu = 2.40 and 2.58.

7. These results could be useful in future policies and PN emission models.

8. Essentially, the superior performance of the battery with the new flow fields can be attributed to improvement in mass transfer and the resulting reduction in concentration polarization losses.

9. Across China, the ratios of total nitrogen to total phosphorus (TN: TP) were high (> 16) in most rivers (93.88%) and had a mean value of 58.24:1 (supplementary Fig. S3), considerably higher than the Redfield ratio of 16:1 (Redfield, 1958).

10. Results show that, with increasing training data, the model loss for either y1 or y2 decreases rapidly and gradually keeps stable, proving that the volume of data collected in this study is sufficient to train CNNs with the optimal performance.

Task 6.5 Here are the sentences of a research paper used to comment on the results. Work with a partner to put them back in the correct order.

A. Particularly, with increasing the flow rate, the difference between the SFF and the other four flow fields becomes increasingly significant, primarily because the pressure loss of the flow channel is positively associated with the local electrolyte velocity.

B. Fig. 21 compares the pressure drop of electrolyte flow in the RFB with different flow fields (at inlet flow rates from 35 to 75 mL · min^{-1}).

C. It is found that all the three new flow fields yield comparable pressure drop with the IFF while exhibiting broadly lower pressure drop as compared with the SFF.

Task 6.6 Read the excerpt from the *Results* and *Discussion* section of Sample Article 5 (SA5) and match the sentences with the information elements in the box.

A. Reviewing the method

B. Summarize major findings

C. Highlighting specific results

D. Giving reasons for the results

E. Assessing the results

F. Indicating graphics

G. Interpreting the results

Can Masks Be Reused After Hot Water Decontamination During the COVID-19 Pandemic?

Results and discussion

3.5. Effects of actual service processing

1 In order to study the influence of actual service processing on masks, we examined mask samples that had been worn for 8 h by participants. **2** For surgical masks, the effects of wearing varied among individuals and for the same individuals at different times. **3** After being worn for 8 h, followed by hot water decontamination and charge regeneration, the PFE values of the surgical masks decreased by 0.5%–12% in our experiments, based on the testing of 15 samples. **4** However, all of the tested KN95-grade masks (10 samples) that had been worn for 8 h, followed by hot water decontamination and charge regeneration, were able to retain filtration efficiencies greater than 95%. **5** It was observed that the waterproof property of masks that were worn might change, as a result of the adsorption of dirt and oil from the participants' skin. **6** Fig. 5(a) shows droplets on the surface of the outer layer of a typical disposable medical mask under daylight and under a 365 nm UV lamp, respectively. **7** Under UV excitation, the droplets of water lacked brightness but the fluorescent nanodispersion droplets exhibited a bright blue emission due to the fluorescence of the carbon nanodots. **8** After removing the droplets, no significant marks of the fluorescent nanoparticles were observed [Fig. 5(b)], indicating the excellent resistance to wetting of the outer layer of the mask. **9** However, if the wettability of the masks was enhanced by potential interfacial reactions between dirt/oil and the fibers, a spontaneous infiltration process could be promoted. **10** In this work, we developed a fluorescent liquid penetrant test based on fluorescent nanodispersion for investigating masks. **11** To summarize, water-dispersible carbon nanodots [6] were sprayed onto the surface of the outer layer of the mask, and the liquid drops were then allowed to remain for 30 min while the penetrant was sinking into the mask. **12** After that, the excess liquid on the surface of the outer layer of the mask was removed, and the inner layer of the mask was inspected under the irradiation of a hand-held UV lamp with emission at 365 nm. **13** As shown in Fig. 5(c), no significant fluorescence signal of the nanodots was observed when testing new masks. **14** However, the fluorescent nanoparticles penetrated into the surface defects of a mask that was damaged by oil, with the fluorescence signal mainly being concentrated in the "damaged area," where the hydrophobicity of the fibers changed. **15** These findings may aid in the development of new materials and design strategies for reusable masks.

Evaluating the study

This information chunk moves readers' attention away from the specific results of the study and begins to focus more generally on the importance that the overall study may have for relevant fields. Authors may evaluate a study from different perspectives. The following two exemplify the common elements that may be contained in this chunk.

IE8: Discussing the strengths of the study

It is important for authors, when writing the paper, to give their readers good reasons why their study is valuable and worth publishing. A common practice is to explicitly present the strengths of the study over existing studies on the same topic. When highlighting the significance of your study, you don't want to appear arrogant, condescending or patronizing. You should not boast or exaggerate. Meanwhile, avoid criticizing other studies bluntly or sounding offensive to show respect for other researchers. Below are some examples.

Several previous studies have used economic models to determine the optimal price of electricity generated by wind and solar power plants [91,98,99]. In this study, the multi-criteria spatial decision-making system and economic model are combined to achieve a more accurate model.

IE9: Discussing the limitations of the study

Of course, no research could be designed and performed perfectly. Almost all research projects have certain limitations. Do not dodge talking about limitations. Objectively discussing the limitations shows researchers' awareness of reflecting on the possible issues of their work. When addressing the issues, you can explain their reasons or effects on the study, or mention the plan to deal with them. In this way, you can enhance the reliability and validity of the study. See some examples below.

The limitations of the present study are: (i) the lack of accessibility to appropriate data for mapping some of the criteria such as bird habitats and marshlands, (ii) uncertainty in the expert opinions, and (iii) determination of optimal values of the parameters used in the NPV model in different geographical areas.

Although promising, these existing methods have at least one of several limitations: extend period of functional loss during self-repair; need for manual reassembly; or need for external energy sources such as heat, light or mechanical energy. Overcoming these limitations has the potential to dramatically improve the longevity, performance and functionality of deformable electronic materials.

Task 6.7 Read the following sentences and underline the words or expressions indicating limitations. Please summarize two ways to convey limitations.

1. However, it is still difficult to interpret the spatiotemporal characteristics of planning implementation.

2. There are some limitations in this study that we hope can be addressed in the future.

3. One flaw associated with this method is the simplified assumption that the nearest park is the park being visited most, which is not always the case.

4. This has the disadvantage that the growth substrate is often highly variable, reducing the replicability of design.

5. This approach has the weakness that any perceived patterns and trends cannot be validated by statistical methods.

6. Our data showed no significant influence of soil saturation on tree stability, but the experimental design did not allow us to fully evaluate this relationship.

Task 6.8 The following text is the excerpt from the *Discussion* section of a research article. Read it carefully and answer the questions that follow.

1 In this study we did not consider the effect of vacancy concentration on the total diffusivity. **2** The vacancy concentration is an important parameter, which also depends on the alloy composition through the vacancy formation energy. **3** Preliminary *ab initio* and classical MD calculations demonstrated that the vacancy formation energies in the NiFe equiatomic alloy and other concentrated solid solutions [10,48] vary within a wide

> range depending on the local atomic configurations. **4** However, detailed investigations of these effects have not yet been performed and accurate estimations of the concentration of mobile vacancies in concentrated alloys with chemically different components are currently impossible. **5** For this reason, in this study we limited ourselves to the effects of composition on a single vacancy migration.

1. What is the limitation of the study?

2. There are two ways to convey limitations (see Task 6.7). How do the authors convey the limitation?

3. How do the authors lessen the negative impact of the limitation?

4. Do you think the limitation is acceptable? If not, why not? If so, why?

5. Do you think it would be a good idea if the author did not discuss the limitations? Why or why not?

6. Does discussing the limitations signal a failure of the study/researcher? Give your reasons.

Concluding the study

> This information chunk signals the ending of the paper. Authors may have their own choices of what to put in this final part, while some common information elements are concluding the research, talking about the values/contributions of the study or the potential implications/applications of the findings, and making recommendations for future research. Sometimes, this chunk of information may stand alone to form an independent *Conclusion* section. In such cases, authors may choose to initiate this last section by reviewing the research (IE1) and summarizing major findings (IE2), or by just continuing their discussion in the preceding section.
>
> **IE10: Making generalizations**
> At this final stage, authors may take the last chance to strengthen readers' impression of what their study has detected and fulfilled, and particularly what generalizations are

drawn from their work. This element sometimes is similar to IE2 — summarizing major findings. A possible way to distinguish the two could be the tense used in the sentences. The summary of the results is usually expressed in the simple past tense because it is to present what was found in the finished work; whereas the simple present tense, often containing modal auxiliaries, is commonly used when expressing generalized understandings drawn from what was observed or detected from the research procedures. See the examples below.

In conclusion, hydrogen addition can clearly improve the engine performance but increase NOx emissions.

The optimal feature of the very suitable locations ensures the highest benefit for the investor and consequently overpayment on the government's part for purchasing electricity from wind-power plants in these locations.

IE11: Discussing implications/applications
A research project is often motivated by the need to solve a theoretical or practical problem. At the end of the paper, it will be a good practice to address this need by talking about any implications/applications of the study findings or the theoretical and/or practical values/contributions that the study may have for relevant fields. See the examples below.

We believe that on the basis of the ultra-flexible 3D-TENG, many kinds of application areas will spring out to make people's life more fascinating.

Models with a location-attention mechanism would be a promising supplementary diagnostic method for frontline clinical doctors.

This element gives authors one more chance to promote their work. However, authors should choose words carefully and avoid clichés and over-generalized expressions, such as "*The present study has greatly contributed to both theoretical development and experimental practice.*"

IE12: Making recommendations
This element allows authors to offer recommendations for the practical applications of

the study findings or the course of future research by pinpointing particular research questions to be addressed or improvements to be made regarding the research methodology. See the examples below.

Future investigations are required to finetune the flow rates of media and air toward the development of SoCE with better differentiation and barrier functions similar to normal human skin.

In the future, integration of the organ-on-chip device with a fluidic control system, an on-stage incubator, and a downstream fraction collector will enable the automation of culture and testing processes.

Task 6.9 The following sentences are abridged from the *Discussion (& Conclusion)* sections. Match the sentences with the information elements listed in the box.

A. Discussing implications   C. Discussing applications

B. Making generalizations   D. Making recommendations

1. The discovery of the design rules can contribute to a deeper understanding of the problem of designing flow fields for RFBs. Furthermore, since the geometric properties are easy to compute, the design rules are believed useful for future research and development.

2. Multi-objective optimization problems will be focused on in future research, including improving the battery's durability and optimization of component sizing and energy management simultaneously.

3. We hope that this short communication article can contribute to alleviating the panic over mask shortage and to promoting new methods for the detection of mask damage zones and the optimal design of reusable masks.

4. One important potential application of our method is brain imaging, because the response of fluorescent contrast agents due to neurons or groups of neurons firing may follow a model similar to (8) [42], [43].

5. Organic and inorganic carbon cycles in rivers are interrelated, and this study implies the need to dynamically estimate riverine greenhouse gas emissions, sediment carbon burial, and OC export to estuaries under the changing global background.

Task 6.10 Compare the *Conclusion* sections in Sample Articles 1 and 4 (SA1 and SA4). Identify the similarities and differences between them concerning information elements and summarize what information elements are obligatory in the writing of the *Conclusion* section.

A Deep Learning System to Screen Novel Coronavirus Disease 2019 Pneumonia

1 In this multi-center case study, we presented a novel method that can screen CT images of COVID-19 automatically by means of deep learning technologies. **2** Models with a location-attention mechanism can classify COVID-19, IAVP, and healthy cases with an overall accuracy rate of 86.7%, and would be a promising supplementary diagnostic method for frontline clinical doctors.

Machine Learning-Assisted Design of Flow Fields for Redox Flow Batteries

1 In this study, we have developed an end-to-end approach to the design of flow fields for RFBs. **2** A search library of 11 564 flow field designs has been generated by combining a custom-made path generation algorithm, hundreds and thousands of multiphysics simulations, and well-trained CNN regression models. **3** Through a collaborative screening process, eight promising candidates highly different from currently known flow fields for RFBs have been successfully identified. **4** Experimental results have shown that the battery with the newly designed flow fields exhibits around a 22% increase in limiting current density and up to 11% improvement in energy efficiency compared to the conventional SFF. **5** Furthermore, to explore the design rules of flow fields, five types of geometric properties have been proposed to describe the morphological features of flow channels. **6** They are the number of turns (P1), the standard deviated length of straight channels (P2), **7** the orientation bias (P3), the length of saved flow channels (P4), and the torque integral (P5). **8** The identification of the geometric properties shared among the eight lead candidates has revealed the quantitative design rules of flow fields: low P1 (18 1), moderate P2 (2.45 0.16), low P3 (0.23 0.18), high P4 (1490 100), and high P5 (20.1 1.8). **9** The discovery of the design rules can contribute to a deeper understanding of the problem of designing flow fields for RFBs. **10** Furthermore, since the geometric properties are easy to compute, the design rules are believed useful for future research and development.

11 It must be stressed that the methodology developed in this study has broad generality in terms of three aspects. **12** First, this study considers RFB flow fields with a single channel but can readily be extended to flow fields with two or more channels after modifying the path generation algorithm, which will be the focus of our future work. **13** Second, although this study is demonstrated with a laboratory-scale RFB, our design approach will transfer well to scaled-up systems if an appropriate scaling-up method is included in the path generation algorithm (an example is given in Fig. S24 of ESI†). **14** Third, in addition to the application in designing flow fields for RFBs, the methodology is applicable to the investigation and optimization of flow fields in other devices, such as flow fields of fuel cells and cooling plates of lithium-ion battery stacks, as long as reliable simulation tools, accumulated design intuition, and convenient device fabrication are available.

LANGUAGE CONVENTION

Read the following sentences and answer the questions that follow.

1. First, it uses fewer pairwise comparisons, and second, it has a higher consistency ratio [50,52,81].

2. Compared with approaches that use "high-tech" equipment such as UV radiation towers, hydrogen peroxide vapor generators, or constant-temperature cabinets, our work provides a simpler and more convenient method for the decontamination of masks during the COVID-19 pandemic.

3. Applying renewable energies would allow for better safety measures and stability in energy production, exempt from the adverse climatic effects of using renewable energies.

4. (Findings confirmed the possibility of mismatched operation of the generator and the turbine.) Auxiliary mismatch angle allows changing the generator torque at a constant turbine torque and during transient processes in an autonomous electric power system.

Questions

1. What are the signal expressions used in Sentences 1 and 2 to show the comparison and contrast between the current studies and previous ones?
2. Which tenses are used in Sentences 1 and 2. Can you find some patterns for the use of verb tense in the comparison and contrast of the current findings with previous ones?
3. Why is the modal auxiliary "would" used in Sentence 3? What difference would it make if "would" is removed?
4. What verb tense is used in Sentence 4? Why do you think this tense is used? Would it be a good idea to use a modal auxiliary instead?

Making comparison and contrast

A common and central element in the *Discussion* section is to compare the results in the present study with those in other related works. In fact, making comparisons and contrasts also appears in other places of a research article, for example, in presenting the research results. It is important to use certain linguistic resources, such as comparative structures, signal expressions, and transitional words or phrases, to indicate similarities and/or differences as well as agreement and/or disagreement between various items. The examples below contain multiple instances (the underlined parts) of language used in making comparison and contrast.

In our experiments, we found that most of the regenerated surgical masks had a PFE <u>much higher than</u> 30%, with an average value of 92.3%.

An interesting finding is that some of these candidates have a <u>similar</u> pattern, as highlighted in the <u>same</u> color.

A Common mistake in making a comparison is that the two sides to be compared are not comparable. The examples below showcase this problem and how it can be solved.

Problematic	*There are more students in Class 1 than Class 2.*
Corrected	*There are more students in Class 1 than in Class 2.*
Problematic	*Students in Class 1 are more than Class 2.*
Corrected	*Students in Class 1 are more than those in Class 2.*
Problematic	*Class 1 has more students than Class 2.*
Corrected	*Class 1 has more students than Class 2 does.*

Task 6.11 Fill in the blanks by selecting the appropriate words or expressions listed in the box below.

in comparison	similar	comparison	different
compared	in contrast	similarly	by contrast

1. OC export to estuaries is _____ from that to oceans (Regnier et al., 2013).

2. A _____ was made between models with and without an added location-attention mechanism.

3. Taking the East China Sea for example, most riverine POC is deposited in nearshore estuarine waters (Deng et al., 2006); _____, riverine DOC can diffuse over a relatively long distance (Bai et al., 2013).

4. For 55 rivers in southeastern Asia, Huang et al. (2017) reported a mean DOC/POC ratio of 1.56, _____ to that of European and American rivers (Ludwig et al., 1996).

5. This model achieved an accuracy rate of nearly 94.4% on 6716 cases, and performed _____ on an independent clinical validation set of 1139 cases.

6. _____ with the conventional single-engine system, the achievable minimum fuel consumption of the dual-engine system under DP is 1.9% lower. _____, the dual-engine system's fuel consumption under the CDCS strategy is slightly higher than that of

the single engine system.

7. An experienced radiologist can make judgments on the possibility of COVID-19 based on his or her clinical experience; however, such judgments are easily influenced by subjective factors and individual proficiency. _____, deep-learning system-based screen models reveal more specific and reliable results by digitizing and standardizing the image information. Hence, they can assist physicians in making a clinical decision more accurately.

Reporting verbs

Reporting verbs play an important role in research papers. They are used in quoting, paraphrasing and summarizing to incorporate evidence into one's writing. The correct use of reporting verbs can accurately report the position on or attitude towards the information from the cited sources; effectively help the authors highlight the significance of an idea; and critically evaluate the sources.

Reporting verbs are frequently used in the *Introduction* and *Discussion* sections, but for different functions. In the *Introduction* section, the verbs are used in citing related works to establish the context and identify research gap(s), thus providing justifications for the current study. In the *Discussion* section, they are used for comparing or contrasting the results and findings with those in previous studies to increase the reliability and validity of the study at hand.

Here are some tips on how to use reporting verbs in research writing.

1. Choose the right verbs.

Different reporting verbs perform different functions and embody different attitudes of authors. For example, verbs such as "investigate" and "examine" indicate research activities, whereas "show" and "identify" suggest research findings. As for authors' attitudes, verbs such as "criticize" and "dispute" indicate disagreement, while "confirm" and "support" indicate agreement. Here are more examples of reporting verbs for different functions and attitudes.

Functions	Examples
Talking about research activities	*test, observe, analyze, compare*
Reporting results	*find, show, demonstrate, illustrate*
Interpreting results	*suggest, indicate, show, imply*
Showing attitudes	*agree, disagree, conclude, recommend*
Highlighting viewpoints	*believe, argue, emphasize, stress, speculate*

2. Use appropriate reporting structures.

Reporting verbs are used in some common sentence patterns. The choice of a pattern depends on the grammatical features of the verbs and the context of the information. Here are some examples of different patterns.

Sentence pattern	Reporting verbs	Sample sentence
verb + *that* clause	*argue, claim, conclude, demonstrate, discover, find, maintain, point out, reveal, show, state, suggest*	Based on the adaptive capacity model, Raichlen and Alexander (2017) **suggested that** the relationship between cognition and PA depends on the adaption of the physiological system to stimuli, which emphasizes the role of cognitive stimuli during exercise.
verb + sb./sth. + *for* + noun/gerund	*applaud, blame, condemn, criticize, single out*	The author (2004) **singles out** experiences and genes **for** the formation of psychological traits.
verb + sb./sth. + *as* + noun/gerund	*classify, define, describe, identify, interpret, perceive, refer to, view*	The author (2004) **identifies** experiences and genes **as** two factors for the formation of psychological traits.

3. Use reporting verbs sparingly.

It is essential to use reporting verbs only when necessary. You may also need to use various verbs and structures to make your expressions diversified. Overuse of reporting verbs can make your writing sound repetitive and dull.

Task 6.12 Complete the following sentences with appropriate reporting verbs. For each sentence, several options may be acceptable but for different functions.

1. Johnson and Martin (2005) _____ that modern technology has made academic dishonesty easier to accomplish and harder for the faculty to identify.

2. A 2008 Fudan study _____ that 449 of the students were reported to have been engaged in some form of cheating.

3. As Terry (2010) _____ rising oil prices as a result of the Asian recovery, the estimates in Economic Outlook show that in rich countries oil prices changes have no significant impact on GDP.

4. Swale (2009) _____ that inflation will remain low.

5. It is _____ by Brown et al. (1997) that online learning can enhance classroom learning through rich activities.

Task 6.13 Two reporting verbs are given in each sentence below. Discuss with your partner how the author's attitude would differ by choosing different reporting verbs.

1. The World Health Organization <u>states/argues</u> that improving health is an integral part of improving both economic prosperity and social justice.

2. Faber (2004) <u>believed/warned</u> that any reduction in cabin staffing will compromise safety and comfort, especially on long international flights.

3. Croft (2010) <u>admitted/argued</u> that tuition increases over the past decade have hurt the ability of many students to attend university.

4. The government (2012) <u>assured/declared</u> that radiation levels in the area of the nuclear accident pose an immediate threat to human health.

Using appropriate tenses

In the *Discussion* (& *Conclusion*) section, the choice of verb tense depends on the type of information you want to present.

The first information chunk in this section is reviewing the study that was finished and its findings already obtained. The simple past tense is most commonly used in referring to the research purpose, methods, and findings. See the underlined verbs in the examples below.

The present study proposed an integrated model of GIS-MCE for locating optimal sites for establishing small and large wind power plants and NPV for estimating the optimal purchasing price of electricity generated by wind turbines.

A fluorescent liquid penetrant test based on fluorescent nanodispersion for investigating masks was carried out, and revealed that the adsorption of dirt and oil from participants' skin was partly responsible for changes in the wettability and filterability of polypropylene masks.

When you move into discussing the results or evaluating the study, you may use the past tense to specify which of your results to discuss and what other researchers did or found, but sometimes the simple present tense is preferred in some disciplines or journals. Tenses may also vary in commenting statements. For example, when you explain possible reasons for the findings, your choice of verb tense depends on whether the explanation is restricted to your study (the past tense) or refers to a general condition (the present tense). This is also the case in other information elements, such as comparing with other studies, assessing the results, and discussing strengths or limitations. Modal auxiliaries may be used to emphasize the speculative nature of these statements.

For the KN95-grade masks with good quality, we found that the masks retained a PFE greater than 95% after being treated in pressurized steam at 121 C for 30 min.

It is shown that using the new flow fields brings about up to 22% increase in limiting current density, from 900 to 1100 mA cm^2, indicating a considerable improvement in the transport of active species within porous electrodes.

Finally, you move to conclusions and end the paper with statements addressing contributions or looking into the future. In these sentences, you may use the simple present tense, modal auxiliaries, and tentative verbs.

Models with a location-attention mechanism can classify COVID-19, IAVP, and healthy cases with an overall accuracy rate of 86.7%, and would be a promising supplementary diagnostic method for frontline clinical doctors.

Task 6.14 Underline the verb(s) in each sentence. Discuss with your partner what information element each sentence represents and how verb tenses are used.

1. To explain the phenomena appearing in the spray-wall combustion experiment for heavy-duty diesel engines under low ambient temperature conditions and further improve the success rate of cold start and heat release rate in a diesel engine, the wall-impinging combustion at varied injection pressures with a large aperture nozzle (diameter 0.32 mm) were studied by 0-D and 3-D combustion simulation.

2. The aim of this study was to obtain the optimal combustion and emissions characteristics by the control of hydrogen addition and EGR.

3. Multi-objective optimization problems will be focused on in future research, including improving the battery's durability and optimization of component sizing and energy management simultaneously.

4. We have presented an approach for the fast localization of multiple fluorescent emitters within a highly scattering medium that takes advantage of variations in temporal delays between responses. The method allows formation of super-resolution optical images from highly scattered light.

5. A mathematical model of an electromagnetic CVT was created to control the flexible coupling of the generator and the turbine, which is described by nonlinear differential equations.

6. Results show that, with increasing training data, the model loss for either y_1 or y_2 decreases

rapidly and gradually keeps stable, proving that the volume of data collected in this study is sufficient to train CNNs with the optimal performance.

Negotiating the strength of claims

As has been previously mentioned, reporting verbs and their tenses may be used to show authors' attitudes. For example, in the sentence below, the main verb "demonstrate" shows the author's certainty about the information to show; the present tense of the verb indicates the "trueness" of the information presented in the *that* clause (also in the present tense). Together, these choices indicate that the authors are very confident of the claim they make in this sentence.

*The results **demonstrate** that band gaps **are widened** by confining tin-halide perovskites to thinner sheets separated by organic linker molecules.*

In addition to reporting verbs, there are other ways to show authors' attitudes or make a claim of appropriate strength. One common way is to use modal auxiliaries such as *may, might, could, can, should, ought to, need to, have to and must*. The sentences below showcase how modal verbs are used to modify the strength of a claim.

Claim	Strength
Some of the information may be missed out by the onlooker due to the time gap.	It is possible.
Some of the information must have been missed by the onlooker due to the time gap.	You are virtually certain but do not actually have evidence to prove it.
Some of the information is missed out by the onlooker due to the time gap.	You are certain about it and have sound evidence to support what you argue.

Depending on the context and the level of formality, some other common phrases and strategies used to negotiate the strength of claims may include the following.

Hedging and boosting: Words or phrases such as "possibly," "likely," "seem to," or "obvious" can be used to adjust the strength of a claim. See the examples below.

*As the efficiencies of perovskite solar cells continue to be optimized in the coming years, it **will likely** be necessary to explore new ways to structurally and electronically tune these classes of materials, and also to use such materials together in multifunction architectures.*

*It is **obvious** that larger particles **resulted in** higher separation efficiency.*

Qualifying: You can add conditions or limitations to a claim to make it more precise or accurate. See the example below.

This study found that X is true, but only in certain circumstances.

Counterbalancing: Acknowledging potential counterarguments or alternative explanations to a point can be a way to negotiate the strength of a claim. See the example below.

While X may be a contributing factor, it's important to consider other variables that could also be at play.

Emphasizing evidence: Specific examples or data can be provided to support a claim. See the example below.

This survey shows that 80% of respondents agree with the statement that X is true.

Acknowledging uncertainty: Admitting there is insufficient evidence or a claim is speculative. See the example below.

While there is some evidence to suggest that X is true, more research is needed to confirm this hypothesis.

Task 6.15 Read the following sentences excerpted from the sample articles, and underline the language expressions used to negotiate the strength of the claim.

1. Furthermore, in terms of morphology, a good flow field design should be achieved by satisfying the following three features.

2. These findings may aid in the development of new materials and design strategies for reusable masks.

3. It is found that the SFF has inferior properties: P2 (3.66), P3 (1.00), P4 (640), and P5 (4.45), indicating that the SFF is highly likely to exhibit both a lower uniformity factor and a higher pressure drop, in agreement with the experimental results.

4. It also ensures electromechanical compatibility with other generators in the system.

5. Scientists at 4C Air, Inc. and Stanford University have reported that heat under various levels of humidity is a promising, nondestructive method for the preservation of the filtration properties in N95-grade respirators, since they have found that UV light can potentially impact the material strength and subsequent sealing of masks.

6. Previous studies used data collected near the sea-land interface without a consistent standard (Dai et al., 2012b; Huang et al., 2012; Tian et al., 2013), which might be another reason for the estimation uncertainties (Table S3).

Characteristic expressions

The characteristic expressions used in the *Discussion* section usually perform the following functions:

- Reviewing the research;
- Summarizing research findings;
- Giving reasons for results;
- Interpreting the results;
- Assessing the results;
- Evaluating the study;
- Making recommendations.

Scan the QR code for a list of the expressions. Try to get familiar with them and pick some to use in your future writing.

Task 6.16 Read the following sentences and underline the characteristic expressions you can use in your own writing. Determine the information element each sentence represents.

1. Here, the uniformity factor and pressure drop of the flow fields were calculated using a 3-D multi-physics model (see Section 2.2 and Note S1 of ESI†), which simulates the coupled physicochemical phenomena occurring in the VRFB.

2. Results show that the simulation results agree very well with the experimental data with an average error of lower than 1%, indicating that the calculation results of the model can be regarded as "ground truth" when training machine learning models.

3. Essentially, the superior performance of the battery with the new flow fields can be attributed to improvement in mass transfer and the resulting reduction in concentration polarization losses.

4. An experimental study relative to the pursuit of an application would be the next step that is motivated by this work.

5. We have presented an approach for the fast localization of multiple fluorescent emitters within a highly scattering medium that takes advantage of variations in temporal delays between responses.

6. Moreover, this work provides a design method for controlling atomic transport in CSAs and an effective simulation approach implementing composition-dependent activation energies in kMC simulations, based on DFT calculations of migration barriers.

7. In this study, only fuel consumption is considered in the energy management strategy. Multi-objective optimization problems will be focused on in future research, including improving the battery's durability and optimization of component sizing and energy management simultaneously.

8. One can predict that alloys with the largest differences in the vacancy formation energy between different atomic subsystems at a composition near the site percolation threshold will exhibit the strongest effect.

Check your understanding

Task 6.17 Choose two from the sample articles(SA1–SA12) and read the *Discussion (& Conclusion)* sections. Finish the tasks below.

1. Analyze the IEs included.
2. Locate the IEs not presented in detail and try to explain the reason.
3. Analyze the tenses and voices used.
4. Highlight characteristic expressions.

Task 6.18 Select two journal articles in your field. Analyze the *Discussion (& Conclusion)* sections and finish the tasks below.

1. Read through the *Abstract* of the chosen articles to help understand the *Discussion (& Conclusion)* section better.

2. Read the *Discussion (& Conclusion)* sections carefully by
    - identifying all the IEs;
    - marking verb tenses and characteristic expressions;
    - comparing the two articles in terms of IEs, typical tenses, and characteristic expressions.

3. Summarize your analysis orally or in writing for class discussion next time.

Unit task

Making comparison and contrast

Up to now, you have already learned how to summarize your findings as well as the overall study, how to compare it with other studies, how to interpret, explain or assess your results, how to evaluate the study, and how to draw conclusions and make recommendations. Of multiple ways of discussing your study, comparing your work with other people's works is central and deserves practice. This task aims to help you develop your skills in making comparison and contrast. Do the following to finish this task.

Step 1: Search for three research articles about the same/similar topic.

Step 2: Choose one article as the study at hand and use the other two as references. Compare the studies in terms of the following aspects:
- principal finding(s);
- research methods;
- study results and reasons for the similarities or differences;
- strengths and weaknesses of the study;
- implications or applications of the work.

Step 3: Write a comparison/contrast essay for about 300 words.

Unit 7

Writing the Title and Abstract

Learning objectives

In this unit, you will
- understand the general function and purposes of the *Title* and the *Abstract*;
- learn about the common information elements in the *Abstract*;
- develop linguistic strategies for writing effective titles and abstracts.

Self-evaluation

Answer the following questions based on your knowledge of research articles in your field.
- How long is the *Title* of an article?
- Is the *Title* written in a full sentence or a phrase?
- What information does the *Title* convey?
- How long is the *Abstract* of an article?
- What information is included in the *Abstract*? In what order?

The *Title* and the *Abstract* tend to be written last although they come first. They are always the first or even the only part that people read about the paper. The *Title* is regarded as a precise summary of the *Abstract*, while the *Abstract* is a precise summary of the whole paper. Besides summarizing the main content, the *Title* and the *Abstract* give a first idea of the main contribution(s) of the article. Therefore, readers can figure out the gist of the article and decide whether or not the article is worth reading further without reading the whole paper. Many more people will read the *Title* and the *Abstract* than the whole paper.

This unit will help you understand how to write a concise and attention-drawing title and an effective abstract for your paper.

WRITING THE TITLE

Read the titles in each group and answer the questions that follow.

Group 1

(1) Machine learning-assisted design of flow fields for redox flow batteries

(2) The site selection of wind energy power plant using GIS-multi-criteria evaluation from economic perspectives

(3) A Deep Learning System to Screen Novel Coronavirus Disease 2019 Pneumonia

(4) A numerical investigation of injection pressure effects on wall-impinging ignition at low temperatures for heavy-duty diesel engine

(5) Effect of exhaust gas recirculation and hydrogen direct injection on combustion and emission

(6) Effects of GDI injector deposits on spray and combustion characteristics under different injection conditions

(7) The effects of *n*-pentanol, di-n-butyl ether (DBE) and exhaust gas recirculation on performance and emissions in a compression ignition engine

(8) Simulating a Monitoring System for an Aquaponics Farm

Group 2

(1) New Approach to Environmental Future City Created by ICT: Sustainable City Network

(2) Blockchain Meets IoT: An Architecture for Scalable Access Management in IoT

(3) Comparison of two algorithms for ECG signal denoising: A recurrent neural network and a support vector regression

Group 3

(1) Human activities changed organic carbon transport in Chinese rivers during 2004–2018

(2) Can Masks Be Reused After Hot Water Decontamination During the COVID-19 Pandemic?

Questions:

1. What is the grammatical structure of the titles in Group 1, Group 2, and Group 3 respectively?
2. Which structure is most commonly used in your research field?
3. How are the two parts of the titles in Group 2 linked? What is the main focus of the first and the second part of the titles in Group 2 respectively?
4. Is the pattern of the titles in Group 3 common in your research field? What might be the advantages and disadvantages of this pattern?
5. How many ways could you find to capitalize a title? Please summarize their features. Which capitalizing pattern is simpler in your opinion?
6. Which capitalizing pattern is the most common in your research field?

Types of titles

Titles can be generally classified as indicative or informative. The former indicates what the article is about (e.g., The relationship between A and B), while the latter informs the reader of what the study has found (e.g., A increases B). Whichever type the titles belong to, they often contain two basic information elements:

- the *topic*, i.e., the main, general subject you are writing about;
- the *focus*, i.e., a detailed narrowing down of the topic into the particular area of the research;

Additionally, titles fall into four categories in terms of their grammatical structure.

Nominal phrase titles

Nominal phrase titles consist of nominal phrases, the most commonly observed structure in the writing of RA titles. Of this type, three syntactic structures are identified and respectively named as "uni-head," "bi-head," and "multi-head" nominal phrases. Specifically, the "uni-head" construction is a noun or noun group followed by a post-modifier. The "bi-head" construction contains two nominal groups followed by post-modifiers while the "multi-head" construction contains more than two nominal groups. Below are some examples.

NSL-KDD data set for network-based anomaly detection systems

A laser-based machine vision measurement system for laser forming

Experimental investigations and multi-objective optimization of MQL-assisted milling process for finishing of AISI 4340 steel

Compound titles

Compound titles comprise two parts in succession, which are juxtaposed on either side of (most usually) a colon. The first part points to the topic of the study or provides an intriguing hint of that, while the second part complements the first part by providing supplementary information. See the examples below.

An application of BRANN and MFL methods: Determining the crack type and physical properties on M5 steel sheets

Systematic knowledge-based product redesign: An empirical study of solar power system for an unmanned transport ship

Dynamic prediction for attitude and position of shield machine in tunnelling: A hybrid deep learning method considering dual attention

Referees are not always right: The case of the 3-D graph

Declarative-sentence titles

Declarative-sentence titles contain a statement usually expressing the study results or conclusions. Writers may adopt this type of title to highlight their findings or conclusions and show their confidence in their work. See the examples below.

- Long-term monitoring reveals a highly structured interspecific variability in the climatic control of sporocarp production
- Understanding sea-level change is impossible without both insights from paleo studies and working across disciplines

Interrogative titles

Interrogative titles contain a question introducing the study subject. They are generally more appropriate for titles of editorials, commentaries, and opinion pieces and are rarely used for writing titles of scientific articles.

Task 7.1 Read the titles of some sample articles and complete the table below.

Title	Is the title a noun phrase, a declarative sentence, or a question?	How many words are there in the title?	Can the title be improved?
Machine learning-assisted design of flow fields for redox flow batteries			
The site selection of wind energy power plant using GIS-multi-criteria evaluation from economic perspectives			
Human activities changed organic carbon transport in Chinese rivers during 2004–2018			
Application of electromagnetic continuous variable transmission in hydraulic turbines to increase stability of an off-grid power system			
Can Masks Be Reused After Hot Water Decontamination During the COVID-19 Pandemic?			

Features of a good title

A well-written title concisely and effectively summarizes the content of the paper, catches readers' attention, and enables readers to identify the current paper from other papers in the same research field. In short, a good title should be accurate, concise, and informative. In other words, being informative, concise and accurate are the three essential qualities of a well-written title.

Accuracy

A well-written title should succinctly but precisely reflect the most important information of the paper by including the keywords and cannot mislead readers about the research you are reporting and the conclusion you have drawn from it by improper use of language expressions.

Conciseness

A well-written title yields the largest amount of information about the content with the fewest words. The *Title* is an abstract of the *Abstract*. For this purpose, phrases (noun phrases and gerunds) are often preferred to sentences although complete sentences are increasingly used. It is also advisable to avoid using decorative or empty expressions. Remember redundancy is the cancer of an article.

Informativeness

As a condensed reflection of the paper, the *Title* should provide enough information for readers to predict what could be learned when they finish reading the whole paper. Therefore, the *Title* should be specific but not too general. A Title like *Study of the Aged Population* tells almost nothing about your study. The *Title* should include all information that makes electronic retrieval of the article sensitive and specific.

Task 7.2 Below are the titles of five sample articles. Read them and write down the keywords enabling you to predict the possible contribution of each RA.

1. Machine learning-assisted design of flow fields for redox flow batteries
   Keywords: _____
   Contribution: _____

2. The site selection of wind energy power plant using GIS-multi-criteria evaluation from economic perspectives
   Keywords: _____
   Contribution: _____

3. A Deep Learning System to Screen Novel Coronavirus Disease 2019 Pneumonia
   Keywords: _____
   Contribution: _____

4. Can Masks Be Reused After Hot Water Decontamination During the COVID-19 Pandemic?
   Keywords: _____
   Contribution: _____

5. Human activities changed organic carbon transport in Chinese rivers during 2004–2018
   Keywords: _____
   Contribution: _____

Task 7.3 Complete the following English titles with proper prepositions according to the Chinese titles. After that, read the English titles carefully and list the information you have learned from them.

1. 利用原位 ^{15}N 标记法估算完整土壤 – 植物系统中牧草豆科植物的地下氮总量
   Use _____ *in situ* ^{15}N–labeling to estimate the total below–ground nitrogen of pasture legumes _____ intact soil–plant systems
   Information: _____

2. 干扰和繁殖压力对生物入侵的短期和长期影响

   Short-term and long-term effects _____ disturbance and propagule pressure _____ a biological invasion

   Information: _____

3. 机器学习辅助的氧化还原液流电池流场设计

   Machine learning-assisted design _____ flow fields _____ redox flow batteries

   Information: _____

4. 2004–2018年间人类活动改变了中国河流中的有机碳运输

   Human activities changed organic carbon transport _____ Chinese rivers _____ 2004–2018

   Information: _____

Task 7.4 Compare the two titles in each group. Identify which one is more informative and explain why.

1. (a) Design of a hydraulic system for liquid packaging
   (b) A hydraulic system for liquid packaging

2. (a) The design and development of a system for biomass production and energy balance
   (b) A system for biomass production and energy balance

3. (a) Estimation of forest harvest volume
   (b) Estimation of forest harvest volume on scale extrapolating approach

4. (a) Runoff-sediment response simulation
   (b) Runoff-sediment response simulation in eco-environmental rehabilitation on Loess plateau

5. (a) Distribution of natural Korean pines in Baihe Forestry Bureau based on spatial models
   (b) Distribution of natural Korean pines

How to write an effective title

Generally, the title should indicate answers to some important basic questions such as **what and where**. To write a good title, think about the following questions; the answers to these questions will help you figure out what to include in your title.

- What is my paper about?
- What methods did I use to perform my study?
- What or who was the subject of my study?
- Where was the study carried out, in the laboratory, or the field?
- What were the most important results of my study?

When writing the title, keep the following tips in your mind.

- Think about terms that people would use to search for your study and include them in your title.
- Pick keywords from recent or often-cited titles close to your work. Use clear and specific keywords.
- Use appropriate descriptive words.
- Choose strategically to write the *Title* into a noun phrase, a compound construction or a full sentence so that the information to stress appears near the front.
- Avoid having too many details. Keep the title brief and attractive without omitting key information.
- Delete unnecessary and redundant words.

Task 7.5 The following titles include some redundant information. Identify them and rewrite the titles into concise ones.

1. An investigation into the modelling of telephony data flows

2. A study of a novel system for solving the three-bus problem

3. The development of a CAE tool for the prediction of the steady state and transient behaviour of orbit annular machines

4. An experiment of fracture mechanics study of the combined effect of hydrogen embrittlement and the effect of loss of constraint

5. On the development of a virtual laboratory module for forensic science degree programs

6. An investigation into the long-term effects of a perennial fiber crop, ramie [Boehmeria nivea (L.) Gaud.], on the chemical characteristics and organic matter of soil

7. A recurrent neural network modelling informed by physics for predictive control of nonlinear processes

Task 7.6 Read the following pairs of titles and explain how the two versions are different.

1	Version 1	From isolated actions to systemic transformations: Exploring innovative initiatives on engineering education for sustainable development in Brazil
	Version 2	The exploration of innovative initiatives on engineering education for sustainable development in Brazil—from isolated actions to systemic transformations
2	Version 1	The Monte Carlo computation error of transition probabilities
	Version 2	Computing transition probabilities error achieved by using Monte Carlo methods
3	Version 1	Silicon wafer mechanical strength measurement for surface damage quantification
	Version 2	Quantifying surface damage by measuring the mechanical strength of silicon wafers
4	Version 1	The specification and evaluation of educational software in primary schools
	Version 2	Specifying and evaluating educational software in primary schools
5	Version 1	Control theory for stochastic distributed parameter systems: an engineering perspective
	Version 2	Stochastic distributed parameter systems following the control theory from the perspective of engineering
6	Version 1	An assessment of energy-economic self-sufficient microgrid on the basis of wind turbine, photovoltaic field, wood gasifier, battery, and hydrogen energy storage
	Version 2	Energy-economic assessment of self-sufficient microgrid based on wind turbine, photovoltaic field, wood gasifier, battery, and hydrogen energy storage

WRITING THE ABSTRACT

 Read the *Abstract* of Sample Article 1 (SA1) and specify what function each sentence performs. The first sentence has been given as an example.

Machine learning-assisted design of flow fields for redox flow batteries

ABSTRACT

1 **1** Flow fields are a crucial component of redox flow batteries (RFBs). **2** Conventional flow fields, designed by trial-and-error approaches and limited human intuition, are difficult to optimize, thus limiting the performance of RFBs. **3** Here, we develop an end-to-end approach to the design of flow fields by combining machine learning and experimental methods. **4** A library of 11 564 flow fields is generated using a custom-made path generation algorithm, in which flow fields are elegantly encoded by two-dimensional binary images. **5** To accelerate the discovery process, we train convolutional neural networks with low test errors for predicting the uniformity factor and pressure drop of flow fields (0.59% and 1.37%, respectively). **6** Through a collaborative screening process, eight promising candidates are successfully identified. **7** Experimental validation shows that the battery with the flow fields designed with this approach yields higher electrolyte utilization and exhibits about a 22% increase in limiting current density and up to 11% improvement in energy efficiency compared to the conventional serpentine flow field. **8** Furthermore, statistical analysis suggests that the promising candidates have a saved channel length of 1490 ± 100 and a torque integral of 20.1 ± 1.8, revealing the quantitative design rules of flow fields for the first time.

redox 氧还反应

algorithm 算法
binary 二元的
convolutional neural networks 卷积神经网络

electrolyte 电解质

serpentine flow field 蛇形流场
torque 转矩, 力矩

Sentence No.	Information element
1.	Presenting the background in formation
2.	
3.	
4.	
5.	
6.	
7.	
8.	

Types of *Abstracts*

The *Abstract* is a synopsis of the paper. Along with the title of the research article, the *Abstract* helps readers decide whether the paper is related to their research interests, and whether they wish to read the whole article or not. Besides working as a screening device, the *Abstract* functions as a preview for readers intending to read the whole article, giving them a road map for their reading.

Abstracts fall into two major categories: *indicative* and *informative*.

An *indicative abstract* is usually a single paragraph telling readers what to expect if they read the article; it informs the readers of what the writer will deal with or attempt to prove, rather than a synopsis of the actual results. It helps readers understand the focus, arguments and conclusions of the larger document so that they can determine whether to read it more thoroughly. This type of abstract is more appropriate for review articles or case reports.

In contrast, an *informative abstract* is a condensed version of the article. It is used for more strictly structured documents (like scientific experiments or investigations) and includes the elements of the original research report: its objective, methods, results, and conclusions. It gives a summary of the main factual information, such as the materials and methods, the results and conclusions, etc. This type of abstract is more suited to reports of original research works. It should be written to stand alone; in other words, readers should be able to understand the abstract without reading the entire article. Normally, when writing up research, the informative abstract is better since you give the reader factual information as well as your main opinions.

Task 7.7 Read the abstracts of three sample articles (SA1, SA2 and SA3) and answer the following questions.

1. Complete the table below. What is the average length of an abstract? Is this the case in your field?

Article	How long is the abstract?	Is the abstract indicative or informative?
SA1		
SA3		
SA5		

2. Are there any citations or references to previous research in these abstracts? Is this common in your field?

3. Do the authors of these abstracts use first-person pronouns (*I or we*)? What is the case in your field?

4. A "self-referring" expression, "this study," is used in the abstract of SA3. What is the function of this expression? Is such an expression common in your field?

5. Are acronyms or abbreviations used in these abstracts? Do they occur in your field?

Overall structure

As a miniature of the article, the abstract would include details that allow readers to grasp the main points of the article. It may serve as an "outline" manual. Figure 7.1 presents the common information elements (IEs) in the *Abstract*.

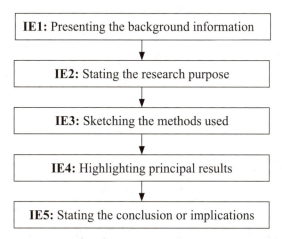

Figure 7.1 Information elements in the *Abstract*

Generally, the elements in an abstract are arranged in the order shown in Fig. 7.1. However, it does not mean that all the elements must appear in an abstract. Some authors may start right by declaring the study objective without mentioning any background information. Some elements may be integrated into one sentence. For example, the method and the aim of the study may be presented in one sentence, or the method and the results may be described together.

Task 7.8 Read the *Abstracts* of Sample Articles 2 and 3. Discuss the questions that follow.

Abstract of SA2

1 The use of wind turbines can help progress towards economic and technological development, lower rates of fossil fuel consumption, decreased greenhouse emissions, and reduced side effects of climate change. **2** A successful mechanism for developing renewable energy worldwide is the guaranteed purchase of electricity generated from renewable energy sources. **3** Accordingly, this study aims to integrate Geographic Information System-Based Multicriteria Evaluation (GIS-MCE) models with economic frameworks to estimate the optimal purchasing price for electricity produced by wind turbines. **4** A total of 13 criteria maps were used and integrated using Ordered Weighted Averaging (OWA) as a type of MCE model. **5** The criteria were initially normalized based on the minimum, and maximum values and weights were assigned to each criterion, using the Best-Worst method. **6** The OWA model identified optimal site locations at various decision risk levels. **7** The economic efficiency of wind turbines and the potential purchasing price of electricity from turbines were also assessed in terms of Net Present Value (NPV). **8** The results show that Ardabil and Southern Khorasan provinces had the most significant areas in the very-suitable class for wind turbine installation (small/large scale). **9** The purchasing prices for wind-generated electricity ranged from 0.047 US$ to 0.182 US$ for large wind farms and 0.074 US$ to 0.384 US$ for small wind plants. **10** The highest electricity produced from large wind farms was found in Maragheh.

Abstract of SA3

1 Rivers serve as regulators of global climate by releasing greenhouse gases, burying particulate carbon, and connecting different ecosystem carbon pools. **2** However, long-term organic carbon (OC) transport features across different Asian rivers are not well known due to unavailable data. **3** Based on routinely monitored environmental and hydrological data during 2004–2018, this study investigated the spatiotemporal variations in dissolved (DOC) and particulate OC (POC) transport across 41 rivers in China. **4** Across different rivers, both DOC (1.35 mg/L–16.8 mg/L) and POC (0.27 mg/L– 4.48 mg/L) concentrations covered wide ranges. **5** The DOC content was high in the north and low in the south, with significantly higher (*t*-test, $p < 0.01$) values for rivers north of 30°N (5.39 ± 3.66 mg/L vs. 2.39 ± 1.14 mg/L). **6** Human activities greatly influenced the riverine DOC and POC distributions. The riverine ammonia nitrogen (NH+ 4-N) content was positively correlated with DOC ($r = 0.81$ and $p < 0.01$) and explained 85.59% of its spatial variation. **7** High vegetation coverage had significant effects on decreasing the riverine POC content, with $r = -0.55$ and $p < 0.05$. **8** During 2004–2018, water pollution prevention and control strategies decreased DOC concentrations in 60.98% of rivers; meanwhile, anthropogenic vegetation restoration and dam construction led to POC content decreases in 90.48% of rivers. **9** Importantly, along with DOC and POC changes, increasing DOC/POC ratios were found in 90.48% of the rivers, with 42.86% being significant, which indicated that Chinese rivers are losing their Asian features of low DOC/POC ratios due to artificial disturbance. **10** This study is significant for accurately quantifying greenhouse gas emissions, carbon burial, and OC export to estuaries by Chinese rivers.

Questions:

1. What information elements are found in the above two abstracts? Complete the following table with the relevant information. The first item has been done as an example.

Article	Background	Research purpose	Method	Results	Contribution/ Implication
SA2	S1, S2				
SA3					

2. Is it possible to present two information elements in one sentence? Could you find such a sentence in the above abstracts?

3. What information elements are important in the above two abstracts? What information elements do the authors focus on in the abstracts? Why?

Task 7.9 Below is a random list of sentences taken from the *Abstract* of a research article. Please identify the information element each sentence represents and rearrange all the sentences in the right order.

A distributed cyber-attack detection scheme with application to DC microgrids

1. _____ The proposed architecture relies on a Lunberger observer together with a bank of Unknown-Input Observers (UIOs) at each subsystem, providing attack detection capabilities.

2. _____ Our analysis shows that some classes of attacks cannot be detected using either module independently; rather, by exploiting both modules simultaneously, we are able to improve the detection properties of the diagnostic tool as a whole.

3. _____ Theoretical results are backed up by simulations, where our method is applied to a realistic model of a low-voltage DC microgrid under attack.

4. _____ We describe the architecture and analyze conditions under which attacks are guaranteed to be detected, and, conversely, when they are stealthy.

5. _____ DC microgrids often present a hierarchical control architecture, requiring the integration of communication layers.

6. _____ Motivated by this application, in this paper we present a distributed monitoring scheme to provide attack-detection capabilities for linear Large-Scale Systems.

7. _____ This leads to the possibility of malicious attackers disrupting the overall system.

LANGUAGE CONVENTION

 Read the *Abstracts* of Sample Articles 1 (SA1) and Sample Article 2 (SA2) again and complete the following table.

IEs	Verb tense	
	SA1	SA2
IE1		
IE2		
IE3		
IE4		

Verb tenses

The verb tenses used in the *Abstract* are similar to the ones used in the corresponding sections of the paper (see Table 7.1). For instance, the opening background and concluding statements are written in the simple present tense or perfect tense, while the methods and results are usually expressed in the past tense. It seems, however, that more abstracts tend to use the simple present.

Table 7.1　Verb tenses in the *Abstract*

Information element	Verb tense
Presenting the background information	simple present tense or present perfect tense
Stating research purpose	simple present tense or simple past tense
Sketching the methods used	simple past tense or simple past tense
Highlighting the principal results	simple past tense or simple past tense
Stating the conclusion or significance	simple present tense or present perfect tense

International Journal Article Writing and Conference Presentation (Science and Engineering)

Task 7.10 Below is the *Abstract* of an RA published in Nano Energy. Please read it carefully and comment on each sentence from the perspectives of language and information elements. The *Title* has been analyzed as an example.

Three-dimensional ultra-flexible triboelectric nanogenerator made by 3D printing	
Excerpt	**Comment**
Title: Three-dimensional ultra-flexible triboelectric nanogenerator made by 3D printing	A noun phrase containing two elements which are linked with "by": ① Topic of the study: Three-dimensional ultra-flexible triboelectric nanogenerator ② Method of the study: 3D printing
Abstract: 1. With the fast developments of wearable devices, artificial intelligence and the Internet of Things, it is important to explore revolutionary approaches and fabrication methods for providing flexible and sustainable power sources.	
2. We report here a practical, ultra-flexible and three-dimensional TENG (3D–TENG) that is capable of driving conventional electronics by harvesting biomechanical energy.	
3. Such TENG is made for the first time by the unique additive manufacturing technology—hybrid UV 3D printing.	

Three-dimensional ultra-flexible triboelectric nanogenerator made by 3D printing	
Excerpt	**Comment**
4. The TENG is made up of printed composite resin parts and ionic hydrogel as the electrification layer and electrode. A sustainable and decent output of 10.98 W/m^3 (P_v, peak power per unit volume) and 0.65 mC/m3 (ρsc, transferred charge per unit volume) are produced under a low triggering frequency of ~ 1.3 Hz, which is attributed to the Maxwell's displacement current.	
5. Meanwhile, a self-powered SOS flickering and buzzing distress signal system and smart lighting shoes are successfully demonstrated, as well as self-powered portable systems of a temperature sensor or a smart watch.	
6. Our work provides new opportunities for constructing multifunctional self-powered systems for applications in realistic environments.	

Characteristic expressions

Characteristic expressions in writing the *Abstract* can be classified into the groups below.

- Stating research purpose
- Sketching the methods used
- Highlighting the principal results
- Stating conclusions or implication

Scan the QR code for a list of the expressions.

Try to get yourself familiar with these expressions and be ready to use them in your future writing.

Check your understanding

Task 7.11 Select three research papers in your field. Finish the tasks below.

1. Compare the titles of the articles in terms of length and structure.

2. Could you find any words or expressions shared by the title and the abstract of each paper? If yes, please underline them.

Task 7.12 Select three research papers in your field. Read the abstracts and fill in the table below.

Abstract	IE(Yes/No)	Sentence (No.)	Tense	First person (Yes/No)
Stating research purpose				
Sketching the methods used				
Highlighting the principal results				
Stating conclusions or implication				

Unit task

Writing the Title and the Abstract

Imagine that you and your team have designed a machine which can remove chewing gum from floors and pavements by treating the gum chemically to transform it into powder and then using vacuum suction to remove it. You have finished writing a paper including the *Introduction*, the *Methods*, the *Results* and the *Discussion*. Only the Discussion is presented below. The other three sections are simply described. Please read them and write an abstract and a title for your paper.

Introduction: In this section, you began by saying that chewing-gum removal is a significant environmental problem. You then provided factual information about the composition of chewing gum [1-2] and the way in which it sticks to the floor.[6] After that, you looked at existing chewing-gum removal machines [3,4] and noted that research has shown that they are unable to use suction to remove gum without damaging the floor surface. You referred to Gumbo et al., who claimed that it was possible to use chemicals to dissolve chewing gum.[5] At the end of the *Introduction*, you announced that you and your research team had designed a chewing gum removal machine (CGRM), which you call GumGone. GumGone sprays a non-toxic chemical onto the gum which transforms it into white powder. The machine can then remove the gum using suction without damaging the floor surface.

Methods: In this section, you described the design and construction of the machine. You compared your CGRM, GumGone, to two existing machines, Gumsucker[3] and Vacu-Gum.[4] You then gave details of a set of trials which you conducted to test the efficiency of the new CGRM and a further set of trials which showed the effect on the floor surface of gum removal.

Results: In this section, you showed the results of these trials. You compared the performance of GumGone with Gumsucker and Vacu-Gum. Your results can be seen in the tables below.

Table 1   Gum removal as a percentage of the total sample

	Gumsucker	**Vacu-Gum**	**GumTone**
Wooden floor	77	73	80
Stone floor	78	78	82
Carpeted floor	56	44	79

Table 2   Floor damage/staining

	Gumsucker	**Vacu-Gum**	**GumTone**
Wooden floor	minimal	minimal	none
Stone floor	significant	some	none
Carpeted floor	significant	significant	minimal

Discussion:

Gum removal technology has traditionally faced the problem of achieving effective gum removal with minimal damage to floor surfaces. Existing CGRMs such as Gumsucker and Vacu-Gum use steam heat and steam injection respectively to remove gum. Although both are fairly effective, the resulting staining and damage to floor surfaces, particularly carpeted floors, is often significant.[10]

In this study, the design and manufacture of a novel CGRM, GumGone, is presented. GumGone reduces the gum to a dry powder using a non-toxic chemical spray and then vacuums the residue, leaving virtually no stain. In trials, GumGone removed a high percentage of gum from all floor surfaces without causing floor damage. The floor surfaces tested included carpeted floors, suggesting that this technology is likely to have considerable commercial use.

Percentage removal levels achieved using GumGone were consistently higher than for existing CGRMs on all types of floor surfaces. This was particularly noticeable in the case of carpeted floors, where 79% of gum was removed from a 400 m^2 area, as opposed to a maximum of 56% with existing machines. This represents a dramatic increase in the percentage amount of gum removed. Our results confirm the theory of Gumbo et al. that chemicals can be used to dissolve gum into dry powder and make it suitable for vacuuming. [5]

The greatest advantage over existing CGRMs, however, lies in the combination of the two technologies in a single machine. The GumGone system resulted in negligible staining of floor surfaces by reducing the delay period between gum treatment and gum removal. This represents a new approach which removes the need for stain treatment or surface repair following gum removal.

As noted earlier, only one wattage level (400 watts of vacuum suction power) was available in the GumGone prototype. Further work is needed to determine the power level at which gum removal is maximized and floor damage remains negligible.

Unit 8

Editing Your Paper

Learning objectives

In this unit, you will
- understand the basic principles of editing manuscripts;
- develop the basic strategies for editing a manuscript.

Self-evaluation

At this stage, you may have completed your manuscript. Think about the following questions.
- Are your ideas clearly and accurately expressed? Have you included unnecessary information?
- Have you provided specific information to allow a clear understanding?
- Is the text easy to follow? Are sentences connected appropriately?
- Do you use formal language? Are there informal words or expressions?
- Are there any language issues? Would you like to have an English teacher or a native English speaker to help polish your paper?

After you finish drafting your manuscript, you need to proofread and edit it to ensure that everything is clear enough and the paper is easy for your fellow scientists to read. You may want to call in an English expert or a native English speaker to improve the manuscript for you, but this is still not the right time. You can, and must, revise and edit your paper by yourself before you relinquish control of the writing process to other people.

Editing is the process of evaluating and making changes to your paper before you submit it to the target journal. This is not an easy job, but it is a crucial part in the course of getting your paper published. Most published works, including books, journal articles, news reports, novels, poems, and songs, have been through several rounds of editing. Good editing can significantly improve the clarity and style of your work.

There are three basic principles of editing your manuscript: conciseness, concreteness, and coherence (3C). In this unit, you will learn how to edit your paper by following these principles.

The 3C EDITING PRINCIPLES

Read the following passage extracted from an *Introduction* section. Discuss with your partner whether it is easy to understand. If not, try to find out how to make it easier to follow by changing some sentences.

1 Nerve injury is one of the most difficult causes to tackle for long-term disability, and regardless of an abundance of studies devoted to the field of regenerative medicine, its treatment options still require further development to date (DiLuca et al., 2014). **2** Nerve injury can be divided into central nerve injury and peripheral nerve injury. **3** Peripheral nerve injury occurs in 1.2% to 2.8% of trauma patients (Carvalho et al., 2019), which is predominantly attributed to traumatic events, such as trauma, joint dislocation, surgical trauma, etc. (Li et al., 2016), may lead to impairment of motor and sensory functions (Yi et al., 2019), further affecting patients' quality of life. **4** The current treatment of choice is tension-free suturing of the peripheral nerve membrane by microsurgery, and in cases where end-to-end suturing of

> the nerve defect is not possible, autologous nerve grafting is the gold standard. **5** Autologous nerve grafting is usually used to bridge long peripheral nerve defects and it is also the "gold standard" in clinical nerve repair (Gu et al., 2014). **6** However, its requirement for secondary operation and the limitation of donor site hinder its clinical application (Tao et al., 2019). **7** While it is challenging to repair, regenerate and recover the function after central nerve injury, it is possible to repair and regenerate injured peripheral nerves.

Editing for conciseness

> Owing to the nature of the topics being discussed, often technical and complex, academic writing should be clear, giving only necessary information, to the point, and describing things objectively. It will be helpful to keep the following in your mind while you are writing your paper.
>
> - Using complex terms does not necessarily make you sound more intelligent. Instead, try to use simple, direct, but clear expressions. Cut everything that is not essential — this will let your key ideas stand out and be identified more easily.
> - Being concise does not always mean using fewer words. It means using the least number of words that make the meaning 100% clear.

Being concise is highly valued in writing, particularly writing for academic purposes. A word from the French writer and poet Antoine de Saint-Exupery may help us appreciate the value of conciseness: "Perfection is achieved, not when there is nothing more to add, but when there is nothing left to take away."

The problem with conciseness may involve both content and language. It can appear within a single sentence or across a range of sentences. Unnecessary repetition of meaning, or redundancy, may confuse your readers and reduce their tendency to read the text in detail. The following sentence showcases this problem.

At this point of time, it now becomes necessary for us to consider alternative possibilities for the purposes of our goals.

This sentence is wordy and unclear, primarily because it is redundant. Some words could be deleted without affecting the meaning of the sentence. For instance, "*at this point of time*" and "*now*" mean the same thing, so the writer could save space by just using one of them. Then, the phrase "*alternative possibilities*" can be simplified into a single word "*alternatives*." In addition, "*for the purposes of our goals*" is unclear in meaning, and it also sounds awkward to use "purposes" and "goals" together. Finally, the writer could use the subject "*we*" to replace "*for us*" and a single verb "*need*" to take the place of "*becomes necessary*." Taking all the modifications together, the sentence will be revised as:

We now need to consider alternatives for achieving our goals.

By cleaning up redundancy, the sentence can be shortened from 21 words to just 10 words without changing its meaning. Compared with the original, the revised version is clearer and much easier to read.

Being concise means you give the information in as few words as possible. However, you may have frequently encountered long sentences in academic texts. Long sentences are indeed needed since research papers often deal with serious topics and convey ideas with high levels of sophistication. However, long sentences often contain multiple layers of ideas expressed in complicated structures and could be hard to follow and even cause confusion. Although you may want to use long sentences to increase the professionalism and formality of your text, you still need to give priority to clarity. Sometimes, you may need to break a long sentence into shorter ones.

Task 8.1 Revise the following text to make it more concise.

1 With the obtained high-resolution structured illumination optical sectioning images, the radius of curvature of two tiger beetles' compound eyes in the red rectangular boxes shown in Fig. 3a, 3e are revealed. **2** Fig. 3b is the maximum intensity projection image and Fig. 3c is the height map.

Task 8.2 Remove the unnecessary repetition in the following texts to improve their conciseness.

1 After obtaining the wet weight, the gastrocnemius muscle of all the groups was stained with Masson staining to assess the recovery of gastrocnemius muscles after sciatic nerve injury at week 16 post-surgery.

2 Previous studies have linked early childhood conduct problems with subsequent drug use. Studies have also found that young adult and adult drug users exhibit more aggressive, uncongenial, and impulsive behaviours than their peers.

1 A total of four design alternatives have been generated by us, and now that the process of their being tested on users has begun in earnest, important and significant information is being learned by our team about which features the users we are concerned with do favour and do not favour. **2** Another piece of crucial information that has been revealed by the testing is the fact that there is great concern among users about the safety of the storage unit. **3** This is interpreted by us to mean not that the current unit is unsafe for users but that the unit designed by us must be at least as safe and, what would be even more preferable, safer for those users. **4** It is for this reason that we intend and plan to incorporate into the unit a feature that completely prevents the lid from collapsing at all when items are being retrieved by users from the interior of the storage unit.

Task 8.3 Revise the following long sentences to make them more concise. The sentences extracted from students' drafts may contain errors. Correct the errors where necessary.

Moreover, it has several applications in the biomedical area with high biocompatibility, controlled biodegradation rate, strong ability to inhibit bacterial growth and good biocompatibility being the main advantages of chitosan.

Following fixation with 2% glutaraldehyde overnight, 1% osmium tetroxide acid was applied to fix and stain the tissues for 2 hours to obtain conduits, then rinsed with gradient alcohol and embedded with an epoxy resin system including dodecenyl succinic anhydride, epoxy resin monomer, nadic methyl anhydride, and 2-4-6-tris (dimethylaminomethyl)-phenol.

Task 8.4 Compare the original text with the revised version. Discuss with your partner what changes have been made and how they improve the text.

Original

1 In assessing the quality of an instrument we distinguish three quality domains, i.e. reliability, validity, and responsiveness. **2** Each domain contains one or more measurement properties. **3** The domain reliability contains three measurement properties: internal consistency, reliability, and measurement error. **4** The domain validity also contains three measurement properties: content validity, construct validity, and criterion validity. **5** The domain responsiveness contains only one measurement property, which is also called responsiveness. **6** The term and definition of the domain and measurement property responsiveness are actually the same, but they are distinguished in the taxonomy for reasons of clarity. **7** Some measurement properties contain one or more aspects, which were defined separately: Content validity includes face validity and construct validity includes structural validity, hypotheses testing, and cross-cultural validity.

Revised

1 We assess each instrument based on reliability, validity, and responsiveness. **2** These domains may be subdivided into measurement properties. **3** Reliability includes internal consistency, reliability, and measurement error; validity includes content validity, construct validity, and criterion validity; responsiveness is both a domain and a measurement property. **4** Some measurement properties contain one or more aspects; for example, content validity includes face validity, and construct validity includes structural validity, hypotheses testing, and cross-cultural validity.

Editing for concreteness

Concreteness in an academic context largely means being specific, i.e., providing specific details for what you are writing about. The details can be facts, examples, statistics, or citations. Sometimes, novice writers err in their papers by not providing such specific details, which makes their writing general and dry. See the example text below.

> The ramp will be reasonably safe. Sheeting with an extremely high coefficient of friction will cover it, making it rather difficult for a wheelchair to slip while

ascending or descending the ramp. Although less expensive sheeting with a lower coefficient of friction could have been used, this seemed ideal to address the needs. In addition, our expert interviews appeared to support it strongly. We did tests under very wet conditions and it performed really well.

In the above example, the authors claim that the ramp is safe due to the use of sheeting with a high friction coefficient. They support the claim by mentioning their interviews and experiments. However, the details as evidence are not specific. For example, how high is the friction coefficient? Who did they interview? How did they carry out the tests? Readers may need more specific details to draw such a conclusion.

The lack of specificity often appears as vague wording. Even if a paragraph is well structured, vague wording can confuse readers and make them doubt your credibility. The paragraph above contains multiple cases of vague wording, which fall into the following categories. Replacement of these vague expressions with precise ones will make your writing more concrete and convincing.

1. Vague modifiers

A modifier qualifies the meaning of a noun or verb. Words such as "reasonably," "extremely," "rather," "difficult," "ideal," "very," "really," and "well" do not qualify the meaning specifically.

Vague   *The ramp will be reasonably safe.*

Precise   *The ramp will keep wheelchairs from slipping in rain or snow.*

2. Vague verbs

The use of verbs, such as "seemed" and "appeared," lower the certainty and specificity of the statements used as evidence to support the authors' claim.

Vague   *Our expert interviews appeared to support it strongly.*

Precise   *Our expert interviews supported it strongly.*

3. Vague pronouns

The use of pronouns may increase textual cohesion, but in academic texts, they are sparingly used. Instead, specific words are used to avoid ambiguity.

Vague	*Our expert interviews supported it.*
Precise	*Our expert interviews confirmed that the sheeting would prevent wheelchairs from slipping on a wet ramp.*

4. Vague references to research methodology

Providing specific details of research methods helps increase the credibility of the claim. In the example text, information about the interviewees, the interview process, and the test procedures would help readers gain a solid understanding of how the conclusion has been drawn.

Vague	*We did tests under very wet conditions.*
Precise	*We covered a mockup of the ramp with the sheeting and sprayed it with a hose for 15 minutes. Then a team member went up and down the ramp in a manual wheelchair.*

5. Vague references to research results

A vague mention of the research results does not inform the readers of what you have found or achieved. It may even impair the credibility of your work.

Vague	*It performed really well.*
Precise	*The team member ascended and descended the ramp three times without the wheelchair slipping at any point.*

Overall, you need to provide specific details to support your claim and avoid using vague expressions in your writing. When you edit your draft, keep asking yourself:

- Have I included specific details to enable a clear understanding of readers?
- Have I provided sound evidence for the credibility of my work?
- Have I had any vague wording which may result in confusion?

Task 8.5 Below is a revised version of the example text analysed above. Discuss with your partner the following questions.

1 Covering the ramp in sheeting keeps wheelchairs from slipping in rain or snow. **2** The sheeting we used is a 20 mesh minus crumb rubber with urethane binders (see Appendix F). **3** The sheeting's high coefficient of friction (0.8) prevents wheelchairs from slipping while ascending or descending the ramp even in wet conditions. **4** Although we could have used less expensive rubber ($2.50 per sq. ft. as opposed to $4.50 per sq. ft.) with a coefficient of friction of 0.6, we chose to use sheeting that conforms to the ADA recommended coefficient of 0.8. **5** Interviews with two Northwestern University professors in the Department of Materials Science confirmed that this sheeting will ensure the safety of users. **6** In addition, our own tests of the material demonstrated its non-slip qualities. **7** We covered a mockup of the ramp with the sheeting and sprayed it with a hose for 15 minutes. **8** Then a team member went up and down the ramp three times in a manual wheelchair. **9** The wheelchair did not slip at any point.

Questions:
1. What revisions have been made?
2. Are the revisions appropriate?
3. Is there any sentence in this revised version that needs further improvement?
4. What changes would you make if you were to revise the example text?

Task 8.6 Read the first paragraph of SA1. Discuss with your partner how the authors avoid vague language expressions and present clear and accurate writing.

1 Developing large-scale energy storage systems is an effective strategy to mitigate power fluctuations of electric grids with a high proportion of renewable energy sources (e.g., solar and wind).[1-3] **2** As one of the most promising energy storage technologies, the redox flow battery (RFB) is attracting increasing attention due to its decoupled energy and power, flexible scalability, fast response, and high safety.[4-6] **3** Among various RFB systems, the all-vanadium redox flow battery (VRFB) is the most widely studied and commercialized as it relies on vanadium ions as both positive and negative electroactive species, which significantly extends the cycle life of the battery by alleviating cross-contamination.[7-9] **4** However, the high capital cost poses a major barrier to the widespread application of the VRFB.[10-12] **5** An effective strategy for cost reduction is to develop VRFBs with higher power density.[13–15] **6** Over the past decades, tremendous progress has been made in increasing reaction sites and improving reaction kinetics of electrodes by surface modification, such as heteroatom doping, catalyst deposition, and surface etching, which are aimed at reducing the activation loss of the battery.[16–18] **7** Another important way to enhance the power density of VRFBs is to decrease the concentration loss by improving electrolyte distribution within the electrode, which can be realized by the design of flow fields.[19–21]

Editing for *coherence*

Generally, the extent to which writing "flows" is referred to as coherence. Coherence is the result of tying information together so that connections you have made in your mind are apparent to the reader. Research papers contain complex technical information that often needs to be expressed in long and complicated sentences, which may increase the difficulty for readers to follow the text. Common solutions to this problem include relating each sentence in the body of a paragraph to the topic sentence and linking each sentence to a previous sentence. In doing so, you make your writing flow smoothly and lead your readers to access and process the intended information as the text unfolds. There are usually four ways to achieve text coherence.

- ***By using connectors that help establish the sequence of events***

Conjunctions or conjunctive adverbs such as *and*, *thus*, and *yet* are widely used to make sentences logically and grammatically more coherent.

- ***By using recycled words***

Repeating some of the words helps link ideas and remind readers of the core information or central idea.

- ***By following the old-to-new pattern***

Start a sentence with information that readers already know, such as knowledge shared by your readers or the elements already mentioned in the previous text. Then, new information is introduced in the rest of the sentence. It will be easier for readers to digest information if they move from what is known (old) to what is to know (new). In the example below, the second sentence begins by repeating the word that appears at the end of the preceding sentence, thus forming a typical old-to-new pattern and a natural transition from sentence 1 to sentence 2.

Populations of co-existing, closely related, but diverging variants of HCV RNA molecules are termed <u>quasispecies</u>. <u>Quasispecies</u> occur in many RNA viruses.

- **By moving from general concepts to increasingly more specific concepts**

Another important aspect regarding text readability is the logical progression within the text. Conventionally, general concepts are provided first, followed by more specific concepts. A view of the big picture will help readers understand specific details. The following example showcases the general-to-specific pattern of text organization.

To keep costs down without sacrificing usability, we have chosen Velcro straps to secure the compartments. Our research revealed that Velcro would cost 75 percent to 90 percent less than our two alternative methods. Furthermore, Velcro is easy to use and durable. Our user observations show that children can easily use Velcro straps to secure and open the compartments. In addition, there are a number of types of Velcro designed for the heavy use the compartments will receive.

Task 8.7 Read the following paragraph. Underline the words and phrases that contribute to the coherence of the text.

1 Medical science has thus succeeded in identifying the hundreds of viruses that can cause the common cold. **2** It has also discovered the most effective means of prevention. **3** One person transmits the cold viruses to another most often by hand. **4** For instance, an infected person covers his mouth to cough. **5** He then picks up the telephone. **6** Half an hour later, his daughter picks up the same telephone. **7** Immediately afterwards, she rubs her eyes. **8** Within a few days, she, too, has a cold. **9** And thus, it spreads. **10** To avoid colds, therefore, people should wash their hands often and keep their hands away from their faces.

Task 8.8 Underline the poorly-written sentence in the paragraph. Discuss with your partner why it is poorly written, and how to improve it.

1 The soil is a major source of pollution. **2** Millions of chemicals are released into the environment and end up in the soil. **3** The impact of most of these chemicals on human health is still not fully known. **4** In addition, in the soil there are naturally occurring amounts of potentially toxic substances whose fate in the terrestrial environment is still poorly known.

Check your understanding

Task 8.9 Compare the two versions of the same text below. Underline the parts that are different. Think carefully about which version you like better and why you like it.

Version 1

1 With the development of new biomaterials, the incorporation of neurotrophic factors and cell transplantation, nerve tissue engineering has provided hope for nerve repair and regeneration. **2** Scaffolds consisting of biological materials can simulate extracellular matrix (ECM) and serve as structural support for endogenous and seeded cells. **3** Thus, the selection of biomaterials becomes a critical factor in nerve tissue engineering. **4** Generally, biological materials include natural and synthetic materials. **5** Natural materials include ECM components (e.g., hyaluronic acid, fibrin, laminin, etc.), chitosan, acellular tissue, silk protein, etc. **6** Although natural materials have good biocompatibility and are suitable for cell adhesion and growth, materials from nonhuman resources acquire certain risks in terms of triggering immune responses and can further damage host tissues. **7** Other disadvantages stem from their heterogeneity and, thus, low reproducibility and repeatability, as well as difficulties in uniform chemical modifications. **8** Conversely, synthetic materials are polymers produced under controlled conditions, which can be further divided into degradable and nondegradable materials. **9** Their main advantages include their easiness of modification according to different needs, presumably less immune response, low toxicity and minimal risk of pathogenic infection. **10** On the other hand, their main disadvantage is poor biocompatibility although this may be improved by certain chemical and biological functionalization.

Version 2

1 The development of nerve tissue engineering with new biomaterials, the incorporation of neurotrophic factors, and cell transplantation has provided hope for nerve repair and regeneration. **2** Scaffolds comprising biological materials can simulate the extracellular matrix (ECM) and serve as structural support for endogenous and seeded cells. **3** Thus, the selection of biomaterials, natural or synthetic, becomes a critical factor in nerve tissue engineering. **4** Natural materials include ECM components (e.g., hyaluronic acid, fibrin, and laminin), chitosan, acellular tissue, and silk protein. **5** Although natural materials have biocompatibility and are suitable for cell adhesion and growth, materials from nonhuman resources can trigger immune responses and further damage host tissues. **6** Other disadvantages of materials from nonhuman resources are their heterogeneity and, thus, low reproducibility and repeatability, as well as difficulties in chemical modification uniformity. **7** Synthetic materials are polymers produced under controlled conditions that can be further divided into degradable and nondegradable materials. **8** The main advantages include their facile modification according to need, presumably less immune response, low toxicity, and minimal risk of pathogenic infection. **9** The main disadvantage of synthetic materials is poor biocompatibility, although this may be improved by chemical and biological functionalization.

Task 8.10 Find a journal article related to your study. Read several paragraphs carefully and analyze the text in terms of conciseness, concreteness, and coherence.

Unit task

Peer Review

After you finish your draft, you may need to have someone else read your paper and give you feedback. This practice is beneficial as the readers may identify problems that you do not realize. In this task, you will work with your partner and review the *Introduction* section for each other. Do the following to finish this task.

Step 1: Checking each paragraph to make sure:
- A topic sentence is upfront showing the focus of the paragraph;
- All the rest sentences are relevant to the focus;
- Ideas are arranged in a general-to-specific pattern;
- Specific information is provided so that readers may have a good understanding and not be confused;
- Sentences are concise and easy to understand;
- Sentences are well connected and information flows naturally and smoothly.

Step 2: Giving comments
- Make comments on your peer's work by using the "Comment" function in MS Word. The comments can be appreciation/agreement on what your peer has done well, and suggestions/recommendations for what could be improved.
- Provide revision suggestions by using the "Track change" function in MS Word. Make sure that your peer can access both the original text and your revisions.

Step 3: Revising your draft
- Revise your draft based on your peer's feedback.
- Use a grammar-checking tool (e.g., Grammarly) to eliminate grammatical errors.

Unit 9

Submitting Your Paper

Learning objectives

In this unit, you will
- understand what are the selection criteria of the target journal;
- learn how to write a cover letter;
- learn how to respond to reviewers' comments.

Self-evaluation

At this stage, you may have completed your manuscript. It is time to think about the following questions concerning how to submit your manuscript.
- What are your target journal's requirements for manuscripts?
- Is it necessary for you to write a cover letter when you submit your manuscript?
- Could you find some specific information about cover letter writing on the homepage of the target journal?
- What is the general principle of responding to reviewers' comments?

Submitting your manuscript to a journal for publication is a long and complicated process that brings you great anxiety and stress. Throughout this process, you may have to negotiate with journal editors and referees back and forth through emails and work hard on revising and editing your manuscript based on their suggestions. Whether your manuscript could be accepted for publication is determined by journal editors and referees using a defined set of selection criteria. Meeting the selection criteria of the journal is thus key to publication success. In this unit, you will learn what are the selection criteria of the target journal, how to write a cover letter, and how to respond to reviewers' comments on your manuscript. All the information will help you develop your publishing strategy and navigate the publishing process, leading to publication success.

SELECTION CRITERIA OF THE TARGET JOURNAL

Select a journal in your field, to which you may submit your paper in the future. Try to find what specific requirements your manuscript should meet before your submission. Take the following three steps to complete this task.

Step 1: Go to the journal website to find author guidelines.
Step 2: Note down the requirements you feel are important or you did not know before reading the guidelines.
Step 3: Exchange what you find with your partners.

A well-organized manuscript will be easier for reviewers to understand and will help them to appreciate its impact. Therefore, the basic selection criterion of a journal is related to the writing style of manuscripts. Most journals provide detailed instructions on how to draft and format papers. Such instructions can be found on the homepage of a journal, often titled as *Author Guidelines*, *Guide for Authors* or *Instructions to Contributors*. Always review the style guide of the journal you select prior to submitting your article.

Target journal's requirements for manuscripts

Task 9.1 The table below is from the Style Guide for Authors of the OSA (The Optical Society of America) journals. Suppose you would submit your manuscript to one of the OSA journals. Discuss with your partners what you should do before your submission.

Elements	Requirements
Title	Informative, accurate, and concise description of the manuscript of your main results
Abstract	• Problem and objectives • Methodology used • Findings and conclusions • Research's effects on broader scientific issues *Note:* Often journals have specific requirements for abstract length and creation.
Introduction	• Problem to be addressed • Background and literature review • New developments and principal results • Research purpose and method
Main Body	• Problem, assumptions, and limitations • Theory and experiment including methods, analysis, derivation, and solution • Results of the study/experiment • Figures and multimedia
Discussion	• Discussion of results and how findings can be viewed in the larger context of the research field • Comparison of results with other related work • Significance of results
Conclusion	• Summary of principal information (no new material introduced) • Statement of specific conclusions • Relevant issues for future consideration
References	• References numbered in order of appearance • Follow the journal's style guide *Note:* Reference management programs can be useful (e.g. EndNote & Bibtex).

Elements	Requirements
Appendices	• Supplementary material for completeness which could detract from the logical presentation of the work • Material that could be valuable to specialists or those wishing to reproduce your results
Acknowledgements	• Technical assistance and useful comments (e.g. from colleagues, advisors, etc.) • Financial support and other relevant disclosures

Task 9.2 Fill in the table below according to the requirements of your target journal. Discuss with your partners any requirements that you think important, confusing, or unexpected. You may discuss questions below.

- How should I balance "be conciseness" and "be informative" in writing my title?
- I have seen many titles starting with "the". Why does the title not begin with an article?
- The words "first," "new," or "novel" seem to claim the novelty of a study. Why should they not be placed at the beginning of the title?

Elements	Requirements
Page layout	
Title	Concise but informative. Not beginning with an article, a preposition, or the words "first," "new," or "novel."
Abstract	
Section headings	
Sections (how many sections are required?)	

Elements	Requirements
References and notes	
Funding	
Acknowledgements	
Disclosures	
Figures	
Supplementary materials	
Tables	
Quality of English	
Video files	
Others	

Reviewers' evaluation

All the submitted RAs must undergo a peer-review process before they are selected for publication in journals. Peer review is a process in which some experts in the field are invited to evaluate the manuscript's quality and suitability for publication. The reviewers should comment on features such as whether the research has been designed, conducted and analyzed in a coherent, valid and ethical way. Their comment not only helps a senior editor decide on whether to accept or reject the paper, but also works as a useful source of feedback, helping writers to improve their paper before it is published. Reviewers' proposal to the senior editor may be: to accept the paper as presented, to

> make changes as recommended by reviewers and repeat the review process, or to reject the paper.
>
> A journal usually takes several months to complete the review process, which typically involves:
>
> - Reading the article and deciding whether to send it for review;
> - Acquiring sufficient reviewers and receiving all feedback;
> - Assessing the reviews and rendering a decision on the paper.

Task 9.3 The following texts are the proposals of some reviewers to the senior editor of a journal. Read them carefully, underline the sentences indicating reviewers' proposals and fill in the blank below each text.

> In the revised manuscript, the authors have highlighted that the generated "perfect" optical vortices are quasi-perfect and not OAM-independent, which has addressed my major concern. Other comments are also well-addressed. I think the manuscript in its present form is suitable for publication in *Optics Letters*.

Reviewer's proposal: _____

> This manuscript demonstrates a volumetric imaging method using ultra-long anti-diffraction (UAD) beams instead of traditional Gaussian beams and Airy beams. Because of the unique curved shape of the UAD beam, it has some potential for future microscopy improvements, such as penetration depth and fast volume scanning. But, I would ask the authors to consider the following comments before I can recommend publication.
>
> - How to analytically describe the so-called ultralong anti-diffracting beam? And the algorithm to convert the 3D spatial beam pattern onto a single-phase pattern on SLM.
> - The UAD images in Fig. 4 and Fig. 5 are very blurry, and both the Gaussian image and the UAD image have some out-of-focus signals. Why and how to solve this problem?
> - The UAD beam proposed is compared against a Gaussian (standard) beam and an Airy beam. It would be appropriate to cite some more relevant literature. For example, Yansheng L. et al, Reports on Progress in Physics, 2020, 83(3) 032401(24). https://doi.org/10.1088/1361-6633/ab7175.

Reviewer's proposal: _____

The authors present an iterative approach and present simulation results of the focus arrays for laser-scanning confocal, STED, and isoSTED microscopy. The topic is interesting and the results demonstrated are valuable to the biomedical optics community. However the writing of the paper has some defects, and some concerns should be addressed and revised according to the following specific comments.

- The description of the intensity distribution near the focal is a little bit simple and superficial. It would be better to provide a more detailed explanation in the physical sense of Eq. (2) and how the Gerchberg-Saxton phase retrieval algorithm worked specifically in the new methods.
- According to the simulation results, the performance of multifocal STED is pretty well. This article would be much more impactful if the experiments of this new approach could be accomplished. Or has this method been experimentally implemented and is there any technical difficulty?
- Are the explanation of the scale bar of Fig.4, 5 and 6 missing?
- The statement of the introduction of the test's specific parameters: "To test our approach, we chose a wavelength of 775 nm, a numerical aperture of 1.35, a refractive index of 1.406, and circular polarization to match typical experimental parameters." probably causes ambiguity. A more clear expression is needed here such as "We stimulated our approach and set the value of wavelength as 755nm…."

Reviewer's proposal: _____

I appreciate that the authors did spend some effort addressing the concerns about the previous manuscript. However, the resolution enhancement they show is unsatisfactory for it is too low for an SR-SIM setup. As we all know that the reconstruction of SR-SIM also employs wiener deconvolution to enhance the spatial resolution, so the resolution of an SR-SIM image should be significantly superior to a Wiener-filtered widefield image. However, the resolution enhancement presented here is indistinguishable from that of conventional deconvolution methods even without requiring any extra optical elements. This disadvantage greatly weakens the significance of this work. It is worth noting that the enhancement in low numerical aperture is less persuasive for SIM communities, especially for high-impact journals like Photonics Research. Thus, I recommend that the authors should improve the filling factor of the high NA system to highlight the significance of this work more persuasively.

Reviewer's proposal: _____

Task 9.4 The following table is the reviewer's report form. Please discuss with your partner on any standards that you think important, or unexpected.

A. GENERAL DATA ON PAPER

Title: _____

_____

_____

No. Of Pages: _____   No. Of Referenes: _____

B. GENERAL EVALUATION OF THE PAPER

	Poor	Below average	Average	Good	Excellent
Theoretical/conceptual framework	☐	☐	☐	☐	☐
Statement of the problem	☐	☐	☐	☐	☐
Significance of research	☐	☐	☐	☐	☐
Literature review	☐	☐	☐	☐	☐
Methodology	☐	☐	☐	☐	☐
Quality of data or findings	☐	☐	☐	☐	☐
Results and conclusion	☐	☐	☐	☐	☐
Readability and writing style	☐	☐	☐	☐	☐

C. ORIGINALITY OF THE PAPER AND ITS CONTRIBUTION TO THE FIELD

	None	Trivial	Modest	Important	Very significant
Contribution to the field	☐	☐	☐	☐	☐

(Source: https://hrcak.srce.hr/upute/guide_reviewers_InterEULawEast_-_Journal_for_International_and_European_Law__Economics_and_Market_Integrations.pdf)

COVER LETTER

 Below is a cover letter. Read it and specify what each paragraph is about.

Dear Editor,

1. I am enclosing here a manuscript of an original research article entitled "Treatment of Hypertension in Patients 80 Years of Age or Order" submitted for consideration of being published in the *New England Journal of Medicine*.

2. Prior to our study, it is unclear whether the treatment of patients with hypertension who are 80 years of age or older is beneficial. Our current study, involving 3,845 patients from Europe, China, Australasia, and Tunisia who were 80 years of age or older, provides evidence that antihypertensive therapy is beneficial and is associated with reduced risks of death from stroke, death from any cause, and heart failure.

3. With the submission of this manuscript, I would like to confirm:
   - All authors of this research paper have contributed to the designing, execution, or analysis of this study;
   - All authors of this paper have read and approved the submitted manuscript;
   - The contents of this manuscript have not been published elsewhere, accepted for publication elsewhere or under editorial review for publication elsewhere;
   - The study was funded by the National Natural Science Foundation of China (70020200). No potential conflict of interest relevant to this article was reported.

4. We deeply appreciate your evaluation and consideration of the paper for publication in your journal and shall look forward to hearing from you at your earliest convenience.

Sincerely Yours,

Peng Wang

Paragraph	What it is about
1	
2	
3	
4	

Contents of a cover letter

Writing a cover letter is an essential part of the journal submission process. A strong cover letter can impact an editor's decision to consider your research paper further and ultimately determine whether to publish it in their journal. A cover letter usually includes the following essential information element.

IE1: Specification of your manuscript

You should first provide basic information about your manuscript: the title and the type of your paper (e.g. original article, review article, case report, communication etc.)

IE2: Summary of the paper to convince the editor

You may briefly describe the rationale and background of your study and the reasons why your paper deserves to be published in the journal. You need to convince the editor that your findings are of significance, theoretically and/or practically.

IE3: Specifying the location

Often in field-based studies, there is a need to describe the study area in great detail. Usually, authors will describe the study region in general terms in the Introduction section and then describe the study site and climate in detail in the Methods section.

IE4: Declaration of conflicts of interest

You may provide information about conflicts of interest that involves any author. You may also include the sources of outside support for research, such as funding and facilities. If there was no conflict of interest, just write: No potential conflict of interest relevant to this article was reported.

IE5: Information about dual submission or prior publication

All journals set policies regarding dual submission or resubmission of previously published papers. You may declare in the cover letter that your paper has not been published elsewhere or submitted to other journals for consideration of publication.

IE6: Inclusion/exclusion of certain reviewers

You may suggest potential reviewers or require that your paper not be reviewed by some people. Provide detailed contact information of the reviewers included or excluded.

IE7: Comment from reviewers of a previous submission to another journal

If you have submitted your manuscript previously to another journal and the manuscript has been reviewed, it is advisable to include the comments you received and any changes or responses you have made based on those comments. This will help to speed up the editorial process.

IE8: Contact information

Include detailed information of the corresponding author (affiliation, mail address, e-mail address, and telephone number).

IE9: Other relevant information

Include any other information you think pertinent and contributive to the editorial process.

Task 9.5 Select an article of your interest from a journal you follow. Pretend that you were the author and would submit the article to the journal. Scan the article to get the necessary information and write a cover letter.

RESPONSE LETTER

 Below is a response letter to the editor's feedback. Read it and specify what each paragraph is about.

Dear Editor,

1 Thank you for your e-mail dated on January 22nd, 2019, informing us that our manuscript (Manuscript ID: XXX) is acceptable for publication in Journal of XXX after revision. We thank the two reviewers for their evaluations and constructive suggestions.

2 We have carefully modified the manuscript to clarify the comments raised by the two reviewers. Please find below our detailed answer to the referees. The major revisions in the revised manuscript are marked in red. We hope that, after the modification and explanation, the manuscript will be suitable for publication in Journal of XXX.

3 Thank you very much for your efforts.

Best regards,
XX (on behalf of all authors)

Paragraph	What it is about
1	
2	
3	

Response to editors' feedback

After weeks of waiting, you will receive the editor's decision letter. There are several possible decisions made by editors.

- **Accept:** The paper is accepted for publication without any further changes required from the authors.

- **Minor Revision:** The paper is accepted for publication in principle once the authors have made some revisions in response to the referees' reports.

- **Major Revision:** A final decision on publication is deferred, pending the authors' response to the referees' comments.

- **Reject and Resubmit:** The paper is rejected because the referees have raised considerable technical objections and/or the authors' claim has not been adequately established. Under these circumstances, the editor's letter will state explicitly whether or not a resubmitted version would be considered.

- **Reject:** The paper is rejected with no offer to reconsider a resubmitted version.

The editor's decision letter usually encloses reviewers' comments, to which you are suggested to respond one by one.

Task 9.6 Below is an editor's e-mail to a writer. Suppose you were the writer and wrote a letter responding to the editor's e-mail.

Dear Professor XXX,

Thank you for submitting your manuscript entitled, "XXX" for publication in Journal of XXX. The review process has now been completed and, guided by the referees' advice, I have decided that the paper should not be accepted for publication in the journal without major revision. Copies of the referees' recommendations are appended below. If you can do this, it would be a great help to me if you would explain your response to the recommendations of the reviewers in the cover letter that accompanies your resubmission.

The journal has a policy that revisions must be submitted within 2 months, or else the paper will be regarded as de facto rejected. It is required that your revision be received by Jan 12, 2023. If you have reasons for not being able to make this deadline, please send an e-mail message as soon as practicable to XX@xj.edu explaining why. You may or may not be granted an extension, based on the substance of the request.

I hope that you and your colleagues will find the referees' comments to be both helpful and constructive. I also hope that you will be willing to revise the manuscript extensively after giving careful consideration to the points that the referees have made.

With kind regards,

Christy Holland, Ph.D.

Editor-in-Chief

Task 9.7 Translate the following editor's reply into Chinese.

Dear Mr. Wang,

Your manuscript has been carefully reviewed. I am returning the reviewers' comments (see the attached file), which I hope you will consider. The reviewers believe that your manuscript is of potential interest to our readers but feel that revision would be necessary before the paper could be considered again for publication in *Surgery*.

Please submit the revision within 60 days of the date of this letter. If you are unable to revise within 60 days, contact us with a request for an extension or your paper will be removed from the editorial process. A request for an extension must be submitted within 60 days of the date of this letter.

I look forward to receiving the revised version of your paper.

Yours sincerely,

Andrew Warshaw

Response to reviewers' comments

A proper response to reviewers' comments on submitted articles is essential to publication. Below is the essential information that should be included in your response letter.

- Title of the manuscript
- A brief "thank you" note addressed to the editor and reviewers stating your gratitude for the review.
- Write responses separately to the comments from different reviewers.
- Format the letter in a way that your responses are distinguished from the reviewers' comments.
- Be sure to answer each and every comment made by the reviewers. This is often called "point-to-point responses to comments."

There are many ways to deal with reviewers' comments, and you will develop your own strategies. Here we outline an approach used by many experienced authors.

- Make all the changes as required or suggested and note each change in your letter of responses.
- Avoid taking a strong or argumentative tone if you happen to disagree on any comments. Instead, state that the reviewer has raised a good point, try to argue in a more positive tone why you do not agree, and provide as many facts as possible to support your argument.

Task 9.8 Below are some reviewers' comments. According to the instructions, respond to the comments properly.

1. **Reviewer's comment:** Page 12, the second paragraph. Authors should explain why they chose hsamiR-520h to examine its role in HSC.

   Response: (agree with the reviewer) _____

   _____

2. **Reviewer's comment:** Discussion Lines 35-40 transfection….functions. Leading to this conclusion the authors have only one major experiment (Fig 3). I think more functional data are needed to validate this hypothesis.

   Response: (not agree with the reviewer) _____

   _____

3. **Reviewer's comment:** In the Introduction section, the 2nd paragraph, line 3: "… over the course of several hundred ….". Probably the word "years" is missing here.

   Response: (agree with the reviewer) _____

   _____

4. **Reviewer's comment:** Please increase the font size of scale bar text and axis labels in figures to make them readable (in print). Also, increase the line strength for the 2D plots in figures.

   Response: (agree with the reviewer) _____

   _____

Check your understanding

Task 9.9 Please exchange ideas with your partner(s) on your manuscript, and try to reply to the comments from your partner(s). You may use the expressions below.

- We thank the reviewer for pointing this out. We have revised….
- We have removed….
- We agree and have updated….
- We have fixed the error.
- This observation is correct….We have changed….
- We have made the change of….The new sentence reads as follows….

Task 9.10 Find a person (one of your peers or professors) who has succeeded in publishing an article in an English journal, ask for all the documents (cover letters, editor's feedback, reviewers' comments, and response to reviewer's comments) incurred in the process, and analyze them with your classmates.

Unit task

Writing a cover letter

After you finish editing your manuscript, you are ready to submit it to the target journal. In this task, you will write a cover letter according to what's offered in this unit. You may also follow the following template to finish the task.

Cover Letter Template

Salutation	Dear Dr./Mr./Ms. [Editor's last name]:
Paragraph 1: Purpose and subject	I am writing to submit our manuscript entitled, [Title] for consideration as a [Journal Name] [Article Type]. [One to two sentence "pitch" that summarizes the study design, where applicable, your research question, your major findings, and the conclusion.]
Paragraph 2: Appropriateness to the journal	Given that [context that prompted your research], we believe that the findings presented in our paper will appeal to the [Reader Profile] who subscribe to [Journal Name]. Our findings will allow your readers to [identify the aspects of the journal's Aim and Scope that align with your paper].
Paragraph 3: Similar works	This manuscript expands on the prior research conducted and published by [Authors] in [Journal Name] or this paper [examines a different aspect of]/ [takes a different approach to] the issues explored in the following papers also published by [Journal Name]. 1. Article 1 2. Article 2 3. Article 3

Paragraph 4: Additional statements often required	Each of the authors confirms that this manuscript has not been previously published and is not currently under consideration by any other journal. Additionally, all of the authors have approved the contents of this paper and have agreed to the [Journal Name] 's submission policies.
Paragraph 5: Potential reviewers	Should you select our manuscript for peer review, we would like to suggest the following potential reviewers/referees because they would have the requisite background to evaluate our findings and interpretation objectively. • [Name, institution, email, expertise] • [Name, institution, email, expertise] • [Name, institution, email, expertise]
Paragraph 6: Frequently requested additional information	Each named author has substantially contributed to conducting the underlying research and drafting this manuscript. Additionally, to the best of our knowledge, the named authors have no conflict of interest, financial or otherwise. Sincerely, [Your Name] Corresponding Author Institution/Affiliation Name [Institution Address] [E-mail address] Additional Contact [should the corresponding author not be available] Institution/Affiliation Name [Institution Address] [E-mail address]

Unit 10

Presenting at Conferences

Learning objectives

In this unit, you will

- understand the functions and principles of the conference communication;
- develop the strategies for delivering oral presentations at conferences;
- practice the skills and techniques of making poster presentations.

Self-evaluation

Scan the QR code to access the poster of the call for papers at an international conference. Please read it and answer the following questions.

- When and where would this conference be held?
- When could the participants register?
- How would participant submit their conference papers?
- What topics would be discussed in the conference (list at least 3 topics)?

Academic conferences, either national or international, online or onsite, have been of great significance in academic exchanges. Researchers attend conferences to publicize their research works and communicate with their peers. To participate in an international conference, you need to write an English research article, correspond with the conference organizer, produce PowerPoint slides and posters conforming to academic norms, deliver a poster presentation or an oral presentation at the conference, and talk with other participants. In addition, you, as a presenter, need to display your work clearly and answer the questions from the audience effectively. As an audience, you need to make appropriate comments on others' statements and communicate or negotiate effectively in discussions. To fulfil these tasks, you should be equipped with some relevant skills and techniques.

In this unit, some activities and tasks have been carefully designed to help you grasp skills and techniques of presenting your research work both visually and orally at an international academic conference.

VISUAL DELIVERY

Look at the following list of elements. How important are they when making a presentation? Discuss their importance in small groups. Are some more important than others in your opinion?

1. **Use your voice effectively**

   Speak clearly, at the right speed and volume with good use of rhythm and pause.

2. **Communicate with correct body language**

   Remain confident yet relaxed and use eye contact and gestures to engage your audience.

3. **Ensure you are prepared**

   Prepare before the presentation so that you feel comfortable to talk using brief notes.

4. **Organize your information**

   Make sure that the content has a clear structure, the parts are linked together, timing has been considered and your talk is relevant and interesting.

5. Handle visual aids professionally

Create relevant diagrams and charts that can be easily read and support the message appropriately and effectively.

6. Make contact with the audience

Establish clear objectives that match audience expectations and involve and stimulate an audience.

Designing presentation slides

Nowadays, it is a common practice to make an academic presentation with a supporting computer-based slide show. A computer slide show allows a presenter to match visual images to an oral presentation. A single image saves a presenter a thousand words. Well-designed slides can greatly increase the retention of details. Most people remember on average about 10 percent of what they hear and 20 percent of what they read. However, with well-designed slides, the retention of an audience that is both hearing and seeing the information can increase to 50 percent.

A well-designed slide for a scientific presentation values clarity and effectiveness. All the elements on the slides should contribute to delivering the intended message. When designing your slides, keep in mind the following tips.

- Use right fonts, sizes and colors.
- Use bullets consistently.
- Avoid wordy expressions and choose the shortest form possible.
- Reduce duplication between the text of the slide and the presentation script.
- Cut brackets containing text.

It is unnecessary to put all examples, definitions or statistics in brackets although you can verbally express them or present some pictures instead. In this way, you can also effectively reduce the duplication between what you show on the slides and what you say to your audience. Below are some examples.

Original slide	Alternative energy (solar, wind, nuclear, etc.)
Edited slide	Alternative energy
Scripts	Today we are going to explore such energies as solar energy, wind energy, and even nuclear energy.

Task 10.1 The following table lists some typestyles. Please tick the typestyles appropriate for presentation slides and elaborate on your choice.

Type style	Example	
Times New Roman	A body in motion will remain in motion.	☐
Bauhaus 93	A body in motion will remain in motion.	☐
Arial black	A body in motion will remain in motion.	☐
Comic Sans MS	A body in motion will remain in motion.	☐
Arial	A body in motion will remain in motion.	☐
Forte	A body in motion will remain in motion.	☐

Task 10.2 Analyze the following slides and try to find the differences between them. Discuss with your partners which one is better designed and explain the possible reasons.

Outline

Introduction
Background
Fillet Design
Computational Results
Experimental Set-up
Experimental Results
Conclusions

Slide 1

This talk presents a computational and experimental analysis of the fillet design.

1. Fillet Design

2. Computational Predictions

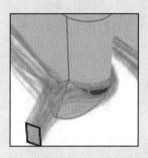

3. Experimental Set-up

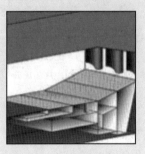

4. Experimental Results

Slide 2

Task 10.3 Try to figure out the problems of the following slides and improve them into appropriate ones.

Discussion

◆ Different optimization goals

- Save storage
- Save CPU utilization

    √ Only if multiple applications are being run together

Modern Mask Optimization Challenges

◆ Large process variation

- Dose and focus variation
- The trade-off between process-variation(PV) band and edge placement error(EPE)

◆ Long OPC runtime

- Take 24 hours on a 1D layer
- Take 3-5 days for metal layers

◆ Spacing of mask patterns cannot be too small

- Resolution limit of e-beam

Task 10.4 Scan the QR code to access the PPT slides of a research proposal. Read them carefully and answer the questions below.

1. What are the major components of this slide presentation?
2. What do you think of the lexical correctness of the texts on the slides?
3. Which part of the slide presentation is heavily loaded with graphics? And why?
4. Are the slides easy to read? Why or why not?

Preparing a poster

Posters form an intrinsic part of scientific life. They may be taken as an advertisement for your research work and yourself in your research community. Posters can be displayed in either printed or electronic format. It is a challenging job to include all the key information within a limited space, for example, with a size of 100cm×200cm. Therefore, designing an effective poster is of great necessity but not an easy job.

The poster design process is composed of planning, selecting, drafting, creating and reviewing steps. Generally, there should be a title, followed by objectives, methods, results, and discussion/conclusion sections. Successful posters are graphically rich presentations of your research that highlight and summarize the main points, with the poster presenter filling in the details in person on-site. The least effective poster, in contrast, is an enlarged copy of your manuscript. Tables and figures are preferably used to allow for a more interesting presentation of the data.

Your poster should include:

- the paper title and all authors at the top;
- a brief introduction, goals, experimental detail, conclusions, and references, presented in a logical and clear sequence;
- tables, graphs, pictures, and explanations for them.

While designing your poster usually according to the templates provided by the conference organizer, you should also learn to create a visually appealing poster, and try to convey your messages using minimal words and strong images.

Some factors should be taken seriously to make posters more eye-catching, such as colours, background images, juxtaposition of texts and graphics, format, and readability.

It is also crucial to check the dimensions of the poster boards in advance. After you have the whole package assembled, you would better print out a draft copy because the computer-generated poster will look very different on the wall. You may invite as many people as you can to give you their feedback.

Full Poster Title Goes Here: And Subtitle if You Have One

Author Name 1, Author Name 2, Author Name 3, Author Name 4

Affiliations

Background

Add text here

- Bullets are easier to read and break up large chunks of text
- Avoid large chunks of text because it makes it harder for the reader to quickly understand your points
- The shorter, the better

Methods

Add text here

- Bullets are easier to read and break up large chunks of text
- Avoid large chunks of text because it makes it harder for the reader to quickly understand your points
- The shorter, the better

Results

Add text here

- Bullets are easier to read and break up large chunks of text
- Avoid large chunks of text because it makes it harder for the reader to quickly understand your points
- The shorter, the better

Add graphics that visually show your results and add interest to your poster

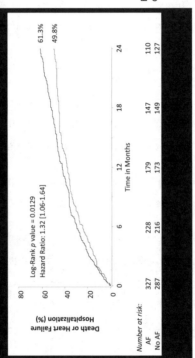

Yale SCHOOL OF MEDICINE

Results

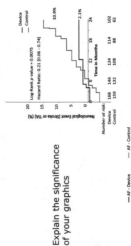

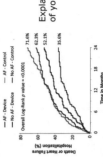

Explain the significance of your graphics

Explain the significance of your graphics

Conclusions and Clinical Implications

Explain the significance of your graphics

Disclosure Information

none

Task 10.5 Scan the QR code to access the sample poster about predictive autofocus technology. Analyze the poster design and comment on it in the table below.

Poster design	Comments
Format	
Text	
Graphics	
Colors	
Readability	

Task 10.6 Select a journal paper of your interest and design a conference poster for the paper. Bring your poster to the class and ask your classmates to comment on it from five aspects, including format, text, graphics, colours and readability.

ORAL DELIVERY

 Read the re-use language expressions below and match the expressions with the particular functions listed in the box.

A. Introducing the affiliation
B. Greeting the audience and introducing the topic
C. Mapping your talk
D. Introducing slides and making transitions
E. Concluding Your talk

Expression	Function
I am a PhD student at….	
I will begin with an introduction to… .Then I will move on to… . After that, I will deal with…. And I will conclude with…	
I'm very grateful to have this opportunity to talk to you about….	
I'd like to take a minute to go over these three take-home points….	
This next slide shows….	

Delivering a paper presentation

We have discussed how to write a research paper in the previous units. Now you should have had a clear idea of what information elements should be included in each section of a research article. However, due to the difference in spoken and written registers, we cannot simply bring a written paper to the conference and read it word by word. The written language is characterized by long complex sentences, sophisticated vocabulary and an impersonal style. In contrast, the spoken language is characterized by short simple sentences, easy vocabulary and a personal style. Shifting the language style is necessary when you are making a conference speech.

Some people may talk with a speech outline, notes or even PowerPoint slides while others may feel safe to write out everything beforehand and deliver the speech with detailed speaking notes at hand. Therefore, it is necessary to understand how to write a speech script.

An oral presentation is usually made up of three parts: introduction, body and conclusion (see Table. 10.1). A successful presentation comes straight to the point, grabbing the attention of the audience at the very beginning and convincing them of the speaker's points with appropriate proof and impressing the audience with a forceful ending.

Table 10.1 The structure of a paper presentation

Section		Slide	Content
Introduction		1	Introducing title, authors, and affiliations
		2–3	Highlighting the motivation of the research Establishing the speaker's credibility Stating the aim of the study
Body	Methods	3^+	Describing the major material and the essential steps of what was done
	Results	2–4	A short presentation of findings using tables or figures
	Discussion	2–4	How does your work fit with previous work? What are the implications of the results? Does your work answer questions raised by others? Does your work raise new questions?
Conclusion		1–2	Making a clear statement of conclusions
			Inviting questions

国际期刊论文写作与会议交流（理工类）
International Journal Article Writing and Conference Presentation (Science and Engineering)

> Due to the time constraints inherent in a standard conference presentation, it is important to convey the most important information concisely. If you have already prepared a manuscript for your work, preparing a talk will become much easier. Usually, delivering three slides per minute is estimated to be an appropriate pace for a scientific presentation. Table 10.1 also shows the approximate number of slides for each part of your oral presentation.

Task 10.7 Please scan the QR code to access four passages abridged from the prologues to oral presentations. Read them carefully and match the passage numbers on the left column with the skills (A–D) that the speakers use.

Passage No.	The skills the speakers use
Passage 1	A. Saying something personal about yourself
Passage 2	B. Asking the audience a question or get them to raise their hands
Passage 3	C. Saying something humorous
Passage 4	D. Getting the audience to imagine situations

Task 10.8 Please scan the QR code to access four passages abridged from the conclusions of oral presentations. Read them carefully and match the passage number on the left column with the skills (A–D) that the speakers use.

Passage No.	The skills the speakers use
Passage 1	A. Summarizing the main points mentioned previously
Passage 2	B. Using interesting statistics
Passage 3	C. Using simple questions to stimulate interest in the audience
Passage 4	D. Using simple language, talking to the audience as if they were a group of friends

Making a poster presentation

When attending the poster session of the conference, you will need to display your printed poster or E-poster. You should check the designated number of your poster, the location for you to display, and the setup time and the duration of your display. Make sure your poster and you appear at the expected time and place. The presenter usually stands with the poster at least during the mandatory time lot but more frequently than this if possible. The popular times for people to view posters are during the poster sessions and also the (lunch) breaks. You need to introduce and explain your work to the viewers. It is a good idea to add a handout to your poster listing the abstract, the key points of your research, and the contact information.

In many conferences, posters will be scored for awards based on the overall presentation, including the abstract, the poster, and the performance of the presenter. Proper dress helps you leave a better impression on the audience. Therefore, attending the conference, you should follow the established dress code for academic occasions. It is always better to get dressed up because you are representing yourself, your research, and your institution.

Task 10.9 Here is the dialogue between the poster presenter and an attendee at the poster session. Please scan the QR code to access the poster and complete the missing information first. Then role-play the conversation by putting forward some other possible questions. Write down these questions you might ask and answer in the box below.

A: Excuse me.
B: Yes?
A: May I introduce myself? I am Dr. Green from California State University.
B: How do you do? What can I do for you?
A: Well, I'd like to ask a couple of questions about your poster.
B: All right.
A: Why did you conduct this research?

B: _____.

A: What did you mainly find?

B: _____.

A: Thank you very much. Now I understand your position much better.

B: Oh, it was a pleasure. I always enjoy meeting people who are interested in my research.

A: Bye!

B: Bye!

Possible questions

Preparing other forms of conference communication

After delivering an oral presentation, presenters are usually faced with a question and answer (Q&A) session, which is another type of impromptu or offhand speaking. In the Q&A session, presenters may get chances to further impress their audiences or may face hard questions and get embarrassed. The key to a successful Q&A session is to know your topic and get well-prepared especially for the common questions that conference presenters often face. For example, audiences probably will further inquire about research basis and methods which are often explained only briefly in a presentation due to the limited time. You may prepare a couple of additional slides displaying some details of those parts and show them to the audience when being asked. This will demonstrate your professional proficiency as well as your confidence in your work.

When answering questions, do not forget to show thanks to the questioners for their attention and interest. You can also ask the questioners to clarify their questions or ask them whether they have further questions. Sometimes you may not be able to answer a question. In such a case you just admit it honestly and skillfully, and may promise to answer it later.

You may also listen to other people's presentations and ask questions as an audience. That way, you need to come up with clear and meaningful questions to get the information you need and present a professional image of yourself to the public. Generally, you can ask questions to clarify points, get additional information, raise different opinions, and get comments or suggestions.

Apart from the Q&A sessions, you may communicate at the reception desk with the receptionists, during the coffee break with attendees from other places, or at a banquet with other participants. You will exchange ideas about the speeches, research, and even travelling plans after the conference. In fact, through talking with other professionals in and out of the meeting, you will get updated with the cutting-edge development in a particular area, get to know new friends and gain insights from others. The casual communication with other experts will help you better enjoy the academic conference.

Task 10.10 Please scan the QR code to access six passages of Questions and Answers in oral presentations. Read them carefully and match the passage numbers on the left column with the skills (A-F) that the speakers use.

Passage No.	The skills the speakers use
Passage 1	A. Avoiding answering the question
Passage 2	B. Changing the topic
Passage 3	C. Answering the question later
Passage 4	D. Leaving the question to others
Passage 5	E. Shifting the attention to another angle of the question
Passage 6	F. Being polite and modest

Task 10.11 Please scan the QR code to access three dialogues recorded at an academic conference. First, role-play each conversation and then match each dialogue with the possible situation in which it could occur.

Dialogue 1	A. At the coffee break
Dialogue 2	B. At the banquet
Dialogue 3	C. At the registration

Characteristic expressions

Have you ever noticed that sometimes the audience feels like the presenter is guiding them on a journey with his or her speech? If yes, what are the secrets of these successful presentations? The secrets lie in a good command of signposts which help to keep the audience's attention and navigate them through the speech. Virtually, a signpost is a verbal statement or visual cue used by the speaker to guide or engage the audience while bringing them through the various stages of a speech or presentation. The statements are used to:

- Begin your presentation;
- Develop your presentation;
- End your presentation.

Scan the QR code for a list of common signposts to deliver a presentation. Try to get yourself familiar with these signposts and be ready to use them in your future presentation.

Task 10.12 Scan the QR code to access the presentation delivered by a winner of the 5-Minute Research Presentation. Please watch it carefully and answer the following questions.

1. How does the speech establish its logical connections?

2. What are the signposts and transitions employed in the speech?

3. How does the speaker employ them effectively?

Check your understanding

Task 10.13 Imagine you are going to deliver an oral presentation at an international conference in your research field. Design a 10-slide presentation using PowerPoint. Decide what to put in your first and last slides. Choose a proper title for each of the rest slides and make sure to remove all the redundancy. Present it to your group and discuss with your teammates whether there are any possible improvements.

Task 10.14 Select one academic speech, watch the video, and give your comments. When assessing the talk, you can refer to the table below.

Factors to consider in the presentation	The presenter tends to do this (good)	Rather than do this (bad)
Core focus	clarifies the main point of the presentation immediately—it is clear to the audience why they should listen	the main point only emerges towards end—not clear where the presentation is going
Structure	each new point is organically connected to the previous point	there are no clear transitions or connections
Pace/Speed	varies pace: speaks slowly for key points, faster for more obvious information	pauses occasionally; maintains the same speed throughout; no pauses
Delivery	sounds natural, enthusiastic, sincere	sounds rather robotic and non-spontaneous

Factors to consider in the presentation	The presenter tends to do this (good)	Rather than do this (bad)
Body language	eyes on audience, moves hands, stands away from the screen, moves around	eyes on screen, PC, ceiling, floor; static, blocks screen
Style	narrative: you want to hear what happened next; lots of personal pronouns and active forms of verbs	technical, passive forms
Language	dynamic, adjectives, very few linkers (also, in addition, moreover, in particular)	very formal, no emotive adjectives, many linkers
Audience / Involvement	involves/entertains the audience—thus maintaining their attention	seems to be talking to him/herself
Text in slides	little or no text	too much text
Graphics	simple graphics or complex graphics built up gradually	complex graphics
Abstract vs. concrete	gives examples	focuses on abstract theory
Statistics	gives counter-intuitive/interesting facts	makes little or no use of facts/ statistics
At the end	you feel inspired/positive	you are indifferent

(The table is adopted from Andrian Wallwork, "Giving an Academic Presentation in English", *English for Academic Research*, https://doi.org/10.1007/978-3-030-95609-7)

Unit task

Research Presentation Competition

Up to now, you have already learned how to deliver your presentation both visually and orally. You have tried to make posters and slides, poster presentations and paper presentations. It is time to set about presenting your own research or research proposal in the **Five-Minute Research Presentation for University Students**. Do the following to finish this task.

Step 1: Scan the QR code to access the competition requirements and read them carefully.

Step 2: Prepare for the competition through the following steps.
- Submit the abstract of the paper;
- Make PPT slides;
- Write a script of your oral presentation;
- Submit a 5-minute video of your research presentation.

Step 3: Study the rubrics of the research presentation competition carefully, and evaluate your performance.

Content (40%)	1. Clarity: Illustrate the design, background, purpose and significance of the speech 2. Integrity: Fulfill the completeness of the speech with IBC (Introduction-Body-Conclusion) structure 3. Appropriateness: Explain concepts or difficult points with information or examples that the audience can understand 4. Logic: Justify the major points of the speech logically
Delivery (30%)	1. Verbal skills: Maintain the attention of the audience by pronunciation and intonation 2. Non-verbal interaction: Interact with the audience using body language, eye contact and other non-verbal communication methods properly 3. Visual aids: Adopt PPT, audio and video clips, pictures and other multimedia aids effectively to facilitate the speech delivery
Language (30%)	1. Accuracy: Use appropriate words and sentence patterns to achieve a certain degree of grammatical accuracy 2. Fluency: Employ cohesive devices to deliver the speech coherently and clearly 3. Complexity: Achieve lexical variety and syntactic complexity to some degree

References

Cargill, M., & O'Connor, P. 2013. Writing scientific research articles: Strategy and steps. John Wiley & Sons.

Boxman, E.S. & Boxman, R.R.L. 2020. Communicating science: A practical guide for engineers and physical Scientist. Beijing: Tsinghua University Press.

Glasman-Dean, H. 2009. Science research writing: For non-native English speakers. London: Imperial College Press.

Kandrashlinda, O.O. & Revina, E.V. 2019. Syntactic features of scientific articles on materials science. Materials Science and Engineering, 483:012024.

McCormack, J., & Slaght, J. 2015. Extended writing & research skills: English for academic study. Beijing: Foreign Language Teaching and Research Press.

Neuen, S. & Tebeaux, E. 2018. Writing science right. Routledge: CPI Group.

Rowe Nicholas. 2017. Academic & Scientific Poster Presentation: A Modern Comprehensive Guide. Berlin: Springer Berlin Heid.

Swales, J.M. & Feak, C.B. 2012. Academic writing for graduate students: Essential tasks and skills(3rd ed.). Ann Arbor: University of Michigan Press.

Wallwork Adrian. 2022. Giving an Academic Presentation in English. Berlin: Springer Berlin Heid.

陈琦, 史文霞. 2021. 通用学术英语写作（高阶）. 西安：西安交通大学出版社.

李芝. 2017. 英语学术论文写作教程. 北京：中国人民大学出版社.

王永祥, 范娜. 2019. 国际学术交流英语. 北京：清华大学出版社.

张智义, 王永祥. 2019. 学术英语写作教程. 北京：清华大学出版社.

Sample Articles

1. Wan, S., Jiang, H., Guo, Z., He, C., Liang, X., & Djilali, N., et al. Machine learning-assisted design of flow fields for redox flow batteries. *Energy & Environmental Science*, 2022, 15: 2874-2888.
2. Shorabeh, S. N., Firozjaei, H. K., Firozjaei, M. K., et al. The site selection of wind energy power plant using GIS-multi-criteria evaluation from economic perspectives. *Renewable and Sustainable Energy Reviews*, 2022, 168:1-21.
3. Liu, D., Tian, L., Jiang, X., Wu, H., et al. Human activities changed organic carbon transport in Chinese rivers during 2004-2018. *Water Research*, 2022, 222:1-12.
4. Xu, X., Jiang, X., Ma, C., et al. A Deep Learning System to Screen Novel Coronavirus Disease 2019 Pneumonia. *Engineering*, 2020, 6:1122–1129.
5. Wang, D., Sun, B., Wang, J., et al. Can Masks Be Reused After Hot Water Decontamination During the COVID-19 Pandemic? *Engineering*, 2020, 6: 1115–1121.
6. Achitaeva, A., Suslov, K., Nazarychev, A., et al. Application of electromagnetic continuous variable transmission in hydraulic turbines to increase stability of an off-grid power system. *Renewable Energy*, 2022, 196: 125-136.
7. Bentz, B., Lin, D., Patel, J., et al. Multiresolution localization with temporal scanning for super-resolution diffuse optical imaging of fluorescence. *IEEE Transactions on Image Processing*, 2020, 29: 830-842.
8. Bo, Y., Liu, F., Wu, H., et al. A numerical investigation of injection pressure effects on wall-impinging ignition at low-temperatures for heavy-duty diesel engine. *Applied Thermal Engineering*, 2021,184: 1-14.
9. Wang, Y., Wang, J., Hao, C., et al. Characteristics of instantaneous particle number (PN) emissions from hybrid electric vehicles under the real-world driving conditions. Fuel, 2021, 286: 1-
10. Osetskya, Y., N., Bélandb, L., K., Barashevd, A., V., et al. On the existence and origin of sluggish diffusion in chemically disordered concentrated alloys. *Current Opinion in Solid State & Materials Science*, 2018, 22: 65–74.
11. Novo, O. Blockchain Meets IoT: An Architecture for Scalable Access Management in IoT. *IEEE Internet of Things Journal*, 2018, 2: 1184-1195.
12. Zhou, Z., Zhang, W., Ma, X., et al. Effects of GDI injector deposits on spray and combustion characteristics under different injection conditions. *Fuel*, 2020, 278: 1-20.